Symmetry
in
Coordination Chemistry

Symmetry in
Coordination Chemistry

JOHN P. FACKLER, JR.
Case Western Reserve University

ACADEMIC PRESS   *New York and London*

Cover design by Jorge Hernandez

ACADEMIC PRESS, INC.
111 Fifth Avenue, New York, New York 10003

United Kingdom Edition published by
ACADEMIC PRESS, INC. (LONDON) LTD.
Berkeley Square House, London W1X 6BA

LIBRARY OF CONGRESS CATALOG CARD NUMBER: 73-154392

PRINTED IN THE UNITED STATES OF AMERICA

To Naomi

Contents

Preface

With the popularization of the symmerty rules designed by R. B. Woodward and Roald Hoffmann to describe certain organic reaction mechanisms it has become increasingly common to present the symmetry properties of molecules in undergraduate organic chemistry courses. In inorganic coordination chemistry, symmetry properties long have been used to provide a simplified description of bonding and structural properties. Since these concepts are so valuable I feel that an elementary knowledge of molecular symmetry should be taught to all chemistry students. With this thought in mind I began to write this book, which will, I hope, bridge the gap between the elementary ideas of bonding and structure learned by freshmen and those more sophisticated concepts used by the practicing chemist. In practice, I have found the material most helpful to supplement the traditional course work of junior–senior inorganic students. It is for them that the problems and examples have been chosen.

I make no apology for excluding some important symmetry-related material in this book. However, I have attempted to present topics which clearly emphasize the use of symmetry in describing the bonding and structure of *transition metal coordination compounds*. I feel that within this framework an early exposure to the beautifully intricate and symmetric "micromolecular world" dealing with molecular structure and bonding can be achieved. This exposure should stimulate the student to think about how steric and electronic properties of molecules can control their chemical reactivity. Once a student begins to think about chemistry in terms of reactions of discrete molecular species with definite shapes, he is well on the way to discovering a whole new, exciting, and esthetically pleasing chemical world.

Acknowledgments

Stimulation for writing this book can be attributed directly to attempts to teach this material to undergraduates. I have taught seminars on symmetry in chemistry to college freshmen and long have felt that the gifted beginning college student can grasp the concepts of symmetry and apply them to chemical problems. In fact, high school science students also could profit from an early exposure to the concepts of symmetry as applied to the molecular world. However, abstract group theory, the underlying mathematical basis for the application of symmetry to chemical problems, usually is presented as such an esoteric subject that even the practicing scientist rarely wishes to devote the time necessary to fully grasp the material. As a result, consequences of symmetry are often grossly neglected.

Certainly this book would not have been written if I had not learned about symmetry and its application to chemical problems as a graduate student working with F. A. Cotton. His "Chemical Applications of Group Theory" [Wiley (Interscience), New York, 1963] remains the best book available for the practicing chemist seeking knowledge of the subject.

Thanks are given to all the various students of mine who have read and criticized the manuscript but especially Charles Cowman and James Smith. Helen Bircher really made the whole thing possible by typing the original manuscript from a poor handwritten copy.

Finally, I would like to thank the International Union of Crystallography for permission to reproduce the figures appearing in the Appendix.

Molecular symmetry and point groups

1

A complete theory of chemical bonding needs to make no assumptions regarding chemical structure. Unfortunately, except for the simplest compounds such as H_2, HD, etc., a detailed theory is not currently available and is unlikely to be fully developed for many years to come. However, by using empirical and semiempirical structural relationships that have evolved over a number of years, we often can correctly predict molecular structures which usually can be confirmed or rejected by spectroscopic techniques. These predicted or experimentally observed structures can provide a basis for meaningful comments about bonding.

One approach to chemical structure starts with the assumption that similar types of ligands produce similarly structured molecules. If, for example, the anion CoI_4^{2-} is tetrahedral, it is reasonable to assume that $CoBr_4^{2-}$ also is tetrahedral. This approach has its obvious limitations. However, additional relationships do occur to the extent that a student acquainted with modern concepts of bonding and structure is likely to guess that $PtCl_4^{2-}$ is planar and $Ni(OH_2)_6^{2+}$ is octahedral. He might even correctly suggest that the fluoride atoms surrounding manganese in K_2NaMnF_6 are at the corners of a *distorted* octahedron. Yet in spite of many recent successes in predicting structures, no one supposed

1

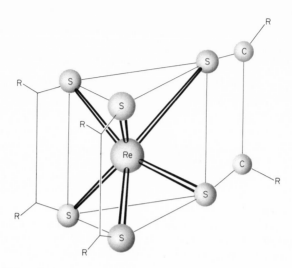

Figure 1-1   *Trigonal prismatic structure as observed for* $Re(S_2C_2R_2)_3$, *where* $R = C_6H_5$.

that $Re(S_2C_2(C_6H_5)_2)_3$ would have a structure (Figure 1-1) in which the six sulfur atoms surrounding the rhenium are arranged at the corners of a trigonal prism. Nor did anyone predict that the anionic dimer $Re_2Cl_8^{2-}$ would have the "cubic" structure pictured in Figure 1-2, and contain a quadruple metal–metal bond. A description of the bonding in such systems has had to await structural determinations.

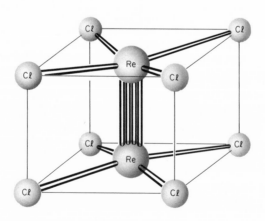

Figure 1-2   *Eclipsed structure of* $Re_2Cl_8^{2-}$.

MOLECULAR SYMMETRY

          The most commonly found structural arrangement for coordination compounds is one in which the groups attached to the central metal atom appear at the corners of an octahedron. Associated with this structure are certain very important symmetry features. There are, for example, three mutually perpendicular axes, x, y, and z, about which a rotation by $n(2\pi/4)$ rad, where $n = 1, 2, 3, \ldots, n$ (90°, 180°, 270°, etc.), makes the molecule indistinguishable from the original species. If the ligands along the x axis are identical to each other, but different from the four identical ligands along y and z, only one such four-fold rotation axis would exist. The octahedral molecule also contains a center of symmetry. This means that each atom has its counterpart at a position with coordinates of identical magnitude and opposite sign (Figure 1-3).
          Since symmetry operations such as rotation, inversion, etc., produce no change in the molecule, any model of bonding we choose also must be invariant with respect to these same symmetry operations.

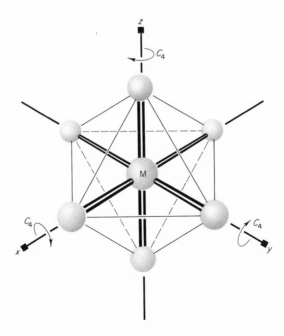

Figure 1-3 *An octahedral complex.*

If, for example, we describe the bond between the metal atom and one of the ligands in an octahedral complex in a particular way, the bonding of the metal to the other five ligands must be described similarly. The bonding model we use must be consistent with the overall structure of the compound. Thus it is apparent that a knowledge of molecular structures can be very important to an understanding of bonding in coordination chemistry—a chemistry noted for its numerous structural types (see Tables 6-3 and 6-4).

Ligand atoms which are appreciably distorted from their free-atom or molecule symmetry by bonding to the metal atom (they are said to be *polarizable*) often produce complexes with nonoctahedral structures. For example, with the polarizable chloride ion, tetrahedral complexes of the type MCl_4^{2-} form for $M = Mn^{II}$, Fe^{II}, Co^{II}, Ni^{II}, Cu^{II}, and Zn^{II}. A trigonal bipyramidal $CuCl_5^{3-}$ also is known. When the metal is bonded to carbon as in cyanides or carbonyls, or to sulfur, phosphorus, arsenic, and several other easily polarized atoms such as metal atoms themselves, unusual structures appear to be the rule rather than the exception. Chemists are just beginning to find out why.'

COORDINATION NUMBERS

In a systematic way we will now explore some of the different symmetries that are observed when varying numbers of ligands are attached or coordinated to one metal ion.

COORDINATION NUMBER TWO

Compounds of the type ML_2 may be linear or bent. In the former case all bonding properties must be the same on rotation by any angle about the molecular axis and on inversion through the center of the molecule (Figure 1-4). With a bent molecule, rotation producing an indistinguishable structure is restricted to $2\pi/2$ rad (180°) about an axis through M and bisecting the L—M—L angle. No center of symmetry exists. Molecularly isolated, condensed-phase (solid or liquid) transition* metal compounds having linear or bent shapes are quite rare. However, some vapor-phase species such as gaseous ZnI_2 are known by structural

---

* Transition elements are defined in Chapter 2.

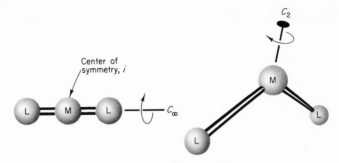

Figure 1-4   *Linear* $D_{\infty h}$ *and bent* C_{2v} *triatomic* ML_2 *compounds.*

studies to be linear. Gaseous MCl_2 species with $M = Mn^{II}$, Fe^{II}, Co^{II}, Ni^{II}, and Cu^{II} are also thought to be linear from spectroscopic studies. Nontransition metal compounds of both structural types are known; H_2S and F_2O are examples of bent compounds, while CO_2 and CS_2 are representative of the linear ML_2 shape.

COORDINATION NUMBER THREE

Either a trigonal planar or a pyramidal shape is expected for ML_3 (Figure 1-5). Both structures have an axis about which a

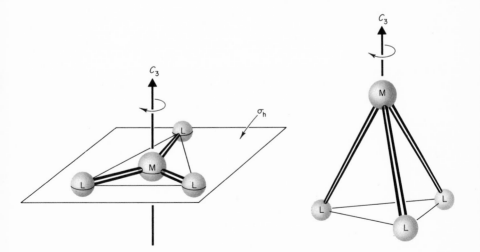

Figure 1-5   *Three-coordinate planar and pyramidal structures.*

rotation by either $2\pi/3$ or $4\pi/3$ (120° or 240°) leads to an equivalent structure. The trigonal planar compound also has a symmetry plane which contains all four atoms. Three-coordinate transition metal compounds also are rather uncommon in the condensed phase. Some complexes of Cu^I, Ag^I, and Hg^{II} are thought to contain essentially a trigonal planar coordination. The species BF_3 and PF_3 are examples of nontransition metal compounds with established planar and pyramidal geometries, respectively.

COORDINATION NUMBER FOUR

Two idealized structures are observed for the coordination number four. These are the tetrahedral and square planar arrangements of ligands (Figures 1-6 and 1-7). Since the tetrahedron can be inscribed in a cube, the structure is called "cubic." This always implies the presence of considerable symmetry. In particular, there are four three-fold rotation axes in every cubic structure (Figure 1-6). These are the principal axes of a tetrahedron. Rotation about these axes by $2\pi/3$ rad or multiples thereof makes the rotated species indistinguishable from the original. There is no center of symmetry in the tetrahedron. Many tetrahedral transition metal compounds are known, especially with anionic halide or

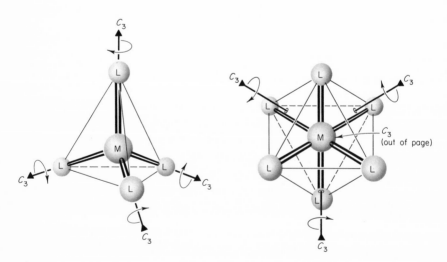

Figure 1-6 *Tetrahedral and octahedral coordination showing the presence of four three-fold axes.*

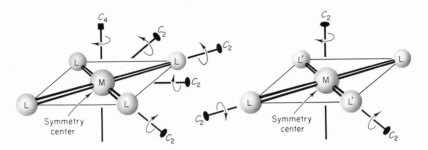

Figure 1-7 *Square planar* ML_4 *and trans planar* ML_2L_2' *coordination.*

pseudohalide ligands, as indicated earlier. Nickel tetracarbonyl, $Ni(CO)_4$, also is tetrahedral.

A planar ML_4 stereochemistry is very commonly observed with bivalent platinum, palladium, copper, and nickel, and nearly always with trivalent gold. Certain chelate* ligands such as

$$\left[\begin{array}{c} \leftarrow\!\!-S\!\!-\!\!C\overset{\displaystyle R}{\diagup} \\ \quad\quad \| \\ \leftarrow\!\!-S\!\!-\!\!C\underset{\displaystyle R}{\diagdown} \end{array} \right]^{2-}$$

tend to promote the formation of planar compounds with many metal ions.

The *square planar*† ML_4 structure is characterized by a principal four-fold rotation axis $(2\pi/4)$ with four two-fold rotation axes perpendicular to it (Figure 1-7). A center of symmetry also exists. Often planar compounds are found in which three two-fold axes (two in the plane, one from C_4), the symmetry plane, and the center are preserved, but the four-fold rotation axis and two two-fold axes of the square are missing. An example is *trans*-$Pt(NH_3)_2Cl_2$, neglecting hydrogen atoms.

---

* The word chelate, χελα, from the Greek, means claw of a crab.

† While "square" requires "planar," the use of "square planar" by chemists is so common that no effort will be made to change this practice here.

Exercise 1-1   A square planar structure can be achieved from a tetrahedral one (and vice versa) by a twist or rotation about the two-fold axis of two ligands relative to the other two ligands. A clockwise twist may lead to a different isomer from that produced by a counterclockwise twist. Consider a tetrahedral Mabcd complex with C_{2a} bisecting angle a—M—b, C_{2c} bisecting c—M—b, and C_{2d} bisecting b—M—d. How many different planar isomers are produced from the two (optical) tetrahedral isomers? Draw them.

COORDINATION NUMBER FIVE

A five-fold rotation axis $(2\pi/5)$ has not been found in a compound in which the metal has a coordination number of five. However, the square pyramidal structure (Figure 1-8) is present in some transition metal compounds. While few transition metal compounds with five identical ligand atoms have this shape,* several compounds of the type MAB_4 are known which approximate it. An X-ray structural determination has shown, for example, that $VO(AcAc)_2$ (Figure 1-9) has nearly a square pyramidal arrangement of the oxygens about the metal. The interaction of planar Cu^{II} compounds with bases such as pyridine, C_5H_5N, may give similarly structured compounds.

A well-known example of a trigonal bipyramidal five-coordinate transition metal compound (Figure 1-10) is iron pentacarbonyl, $Fe(CO)_5$. In this compound, the principal rotation axis is three-fold. In addition to three two-fold axes perpendicular to the three-fold axis, there are four mirror planes, each containing three carbonyls and the iron. The three ligands at the corners of the equilateral triangle perpendicular to the three-fold axis are called *equatorial* ligands; the other two are called *axial*.

COORDINATION NUMBER SIX

In addition to the octahedral structure which is very common, the trigonal prismatic structure (Figure 1-1) is also known. The

---

* One type of $Ni(CN)_5^{3-}$ ion in $[Cr(NH_2CH_2CH_2NH_2)_3][Ni(CN)_5]\cdot1.5\ H_2O$ has a square pyramidal NiL_5 structure, while another is a distorted trigonal bipyramid.

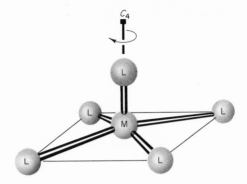

Figure 1-8 *Square pyramidal structure.*

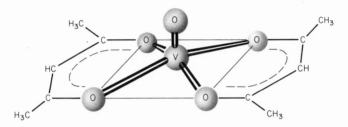

Figure 1-9 *Vanadylacetylacetonate,* VO(AcAc)$_2$.

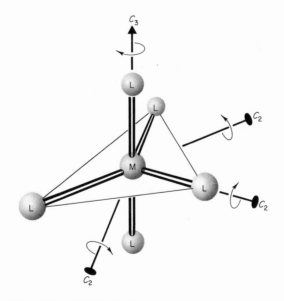

Figure 1-10 *Trigonal bipyramidal structure.*

best example of this arrangement is found in crystalline MoS_2. A few sulfur ligand complexes besides $Re[S_2C_2(C_6H_5)_2]_3$ also are thought to have a similar arrangement of atoms around the metal. In this structure, there are three two-fold rotation axes perpendicular to the three-fold axis; four symmetry planes are also present.

COORDINATION NUMBER EIGHT

Three important structural arrangements have been found for compounds in which the coordination number of the metal atom is eight, with identical ligands. These are the square antiprismatic configuration typified by TaF_8^{3-}, the dodecahedral structure found in $Mo(CN)_8^{3-}$ (Figure 1-11) and the cube. A cubic structure has been observed recently for the anions in Na_3MF_8, M = U, Pa, Np. In the dodecahedral structure there exist three two-fold rotation axes and two mirror planes. In the square antiprismatic configuration the principal rotation axis is four-fold. No inversion center is present in either structure.

The symmetry operations of a cube are identical to those of the octahedron. A tetrahedron displays only one-half of these operations.

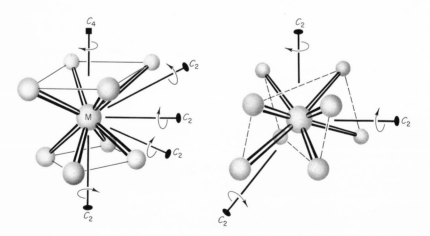

Figure 1-11 *Square antiprismatic and dodecahedral* ML_8 *structures.*

Exercise 1-2   By means of a molecular model, prove to yourself that an inversion center combined with all the symmetry

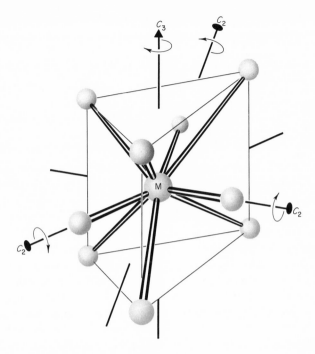

Figure 1-12 *The symmetric, face-centered trigonal prism ML₉ structure of the ReH₉²⁻ anion in K₂ReH₉.*

present in a tetrahedron produces the symmetry operations of an octahedron.

Hint Remember that a tetrahedron can be inscribed in a cube.

COORDINATION NUMBERS SEVEN, NINE, AND HIGHER

Since the number of transition metal compounds having these coordination numbers definitely established is limited [some examples are NbF_7^{2-}, TaF_7^{2-}, and $Nb(H_2O)_9^{5+}$], the various idealized structures observed will not be described here. However, it is apparent that a trigonal prismatic compound containing atoms in the centers of the rectangular faces represents a reasonably high symmetry (Figure 1-12). The anion ReH_9^{2-} has such a structure.

SYMMETRY CLASSIFICATION

As we have observed, the presence of rotation axes, inversion centers, etc. in a molecular structure implies that certain movements of the molecule, *symmetry operations*, produce structurally identical orientations. These symmetry operations are a direct result of the presence of various *symmetry elements*, the rotation axes, inversion center, etc. With the square planar $PtCl_4^{2-}$ anion, the presence of the four-fold rotation axis perpendicular to the plane of the molecule (see Figure 1-7) means that the molecule can be rotated around this axis by 90° $(2\pi/4)$, 180° $(2 \times 2\pi/4)$, 270° $(3 \times 2\pi/4)$, and 360° $(4 \times 2\pi/4)$ to produce an *equivalent* or, in the latter case, an *identical* configuration. This axis is labeled C_4 or simply 4. In general, a C_n or n operation* implies rotation by $2\pi/n$ rad.

In addition to the symmetry elements we have already considered, there is one additional element called an *improper rotation axis*. We speak of n-fold improper rotation axes, S_n. (The proper rotation axis is C_n.) The operation implied by the improper rotation axis is defined as a rotation by $2\pi/n$ rad about the axis followed by reflection through a plane perpendicular to this axis. If an equivalent structure results, an S_n axis is present.

If we examine one of the three-fold rotation axes present in an octahedron (Figure 1-6), we see that it is also a six-fold improper rotation axis, S_6. Rotation about this axis by 60° $(2\pi/6)$ followed by reflection in a (hypothetical) plane perpendicular to this axis produces an equivalent structure. Repeating this process six times brings the figure back to its original configuration. This would not be true, however, if n were odd. In such cases, we must repeat the operation $2n$ times before the structure returns to the original configuration.

Table 1-1 contains a summary of the symmetry elements used to describe molecular structures, together with their required symmetry operations. Two additional symmetry elements are necessary to completely describe three-dimensional crystalline solids, since translational symmetry may be present. To discuss molecular bonding, however, we need consider only those operations which do not change the position of the molecule in space. This means that some point in the molecule, real or imagined, is not moved by any of the symmetry operations associ-

---

* Spectroscopists use C_n, the Schoenflies notation, while crystallographers use the Arabic numeral n, the Hermann–Mauguin notation, to describe an n-fold axis.

Table 1-1 *Symbols (Hermann–Mauguin and Schoenflies) for symmetry elements and their implied operations*

Element	Operation
n, C_n	Rotation by $2\pi/n$ rad
$\bar{1}$, i	Inversion through a symmetry center
S_n	Rotation by $2\pi/n$ rad followed by reflection through a plane perpendicular to S_n (Schoenflies notation only)
$\bar{n}$	Rotation by $2\pi/n$ rad followed by inversion (Hermann–Mauguin notation only)
m, σ	Reflection through the symmetry plane

ated with the molecule—thus, we speak of a point symmetry. In the case of the planar $PtCl_4^{2-}$, this point and the Pt atom coincide. In fact, with many transition metal compounds, the transition metal itself is at this point. When translations are also considered, we speak of space symmetry.

SYMMETRY POINT GROUPS

A collection of elements related by certain specific rules constitutes an *abstract mathematical group*. We will state explicitly what these rules are and leave a detailed discussion of their origin to more advanced texts.*

1. "The product of any two elements in the group and the square of each element must be an element in the group." (CLOSURE)
2. "One element in the group must commute with all the others and leave them unchanged." (IDENTITY)
3. "The associative law of multiplication must hold." (ASSOCIATIVE)
4. "Every element must have a reciprocal which is also an element of the group." (RECIPROCAL)

In classifying molecules into point groups it is necessary to realize that an element as referred to in the above rules means

---

* Since this is a chemistry book, the statements of F. A. Cotton as found in "Chemical Applications of Group Theory," Wiley (Interscience), New York, 1963, have been used.

a symmetry operation—in other words, action. It should not be confused
with a static element of symmetry discussed previously which indicates
some particular symmetry operation or operations such as a C_n axis. The
elements of rule 1 for a group containing only a C_n axis are the rotations
$C_n, C_n^2, C_n^3, C_n^4, \ldots, C_n^{n-1}, C_n^n$. The mathematical terms such as "product,"
"commute," "associative," and "reciprocal" will be familiar to most
readers. However, to avoid confusion they will be used only in the following
manner in this book.

    *The product of the multiplication of elements is the result
obtained by carrying out the implied operations in the order specified.* Hence
the product of rotation by $2\pi/4$ with itself is consecutive rotation twice
by $2\pi/4$ or rotation by π. Thus, $C_4 \times C_4 = C_4^2 = C_2$.

    *The order in which we carry out the operations may be
important since not all operations will commute.* With commuting operations, AB equals BA. However, a reflection followed by rotation will not
always give the same product as a rotation followed by a reflection (unless
the plane of reflection is perpendicular to the rotation axis). Hence these
operations do not always commute, $AB \neq BA$.

    *An associative law means that we can arrange the elements
together in whatever way we choose providing the order is preserved.* Hence
$A \times B \times C \times D$ equals $(A \times B) \times (C \times D)$ or $A \times (B \times C \times D)$, etc.

    In algebra the product of a number with its reciprocal
is one. *For symmetry groups, an operation* (A) *followed by its reciprocal*
(A^{-1}) *leads to a configuration identical to the original.* The resultant element
(symmetry operation) is called the identity, E. All mathematical groups
must have such an element. This element commutes with all others. For
symmetry groups, E is the element implied in rule 2.

MULTIPLICATION OF SYMMETRY OPERATIONS

    To see how the rules of group theory can be applied
to molecules and also to introduce some of the symbolism which is associated with point groups, we consider the symmetry found in the interesting
"sandwich" structure of ferrocene (Figure 1-13). This structure, a twisted
pentagonal prism, contains a five-fold principal rotation axis coincidental
with the axis labeled z. The presence of this C_5 axis requires that rotation
by $m \times (2\pi/5)$, where $m = 2, 3, 4$, and 5, gives configurations indistinguishable from the first one. The last rotation, C_5^5, is symbolized by E,
the identity operation, since it leaves the molecule unchanged. Along with

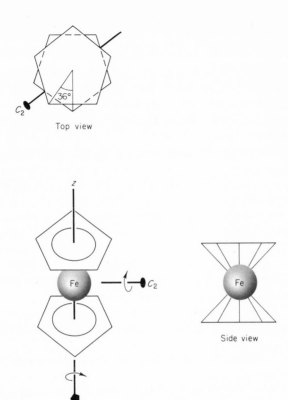

Figure 1-13 *The structure of ferrocene,* $(\pi\text{-}C_5H_5)_2Fe$.

the five-fold rotation axis, z contains a ten-fold improper rotation axis. Rotation by $2\pi/10$ (36°) followed by reflection in a plane perpendicular to this axis carries the molecule into an equivalent, indistinguishable structure. This ten-fold rotation–reflection axis* is labeled S_{10}. As with C_5, various products of S_{10} with itself must lead to indistinguishable configurations. However, some of the $(n-1)$, nine, new operations we might expect to find are not labeled as multiples of S_{10}. For example, "S ten taken five times," S_{10}^{5}, gives the same result that would be obtained by inversion, i. This operation is labeled $\bar{1}$, "bar one" in the crystallographic notation. The presence of the symmetry element requires the operation and vice

---

* Or it is a five-fold rotation–inversion axis, $\bar{5}$ ("bar five").

versa. We will use the rotation–reflection terminology throughout most of this text, but the reader should be able to work with either system.

Parallel to the C_5 (or S_{10}) axis in ferrocene are five mirror planes. The presence of any one such plane containing a C_n axis requires that $(n - 1)$ additional symmetry planes also contain this axis. These n planes are labeled σ_v or σ_d, depending on whether the planes contain two-fold axes perpendicular to C_n (σ_v) or bisect the angle between such axes (σ_d). Rotation of a σ_v plane by C_n will produce the other σ_v planes. If σ_d planes are present, rotation by C_n will generate each of these from an initial one. Only one set of vertical planes is present in ferrocene (Figure 1-13). These are σ_d planes, since they bisect the angle between the C_2 axes. In planar $PtCl_4^{2-}$ (Figure 1-7) both σ_v and σ_d planes are present. A plane perpendicular to the principal rotation axis is given the special symbol σ_h (horizontal). No such plane is present in ferrocene, but there is one in $PtCl_4^{2-}$.

The presence of one two-fold axis perpendicular to the C_5 axis in ferrocene requires that four more C_2 axes be present. Rotation by $2\pi/5$ about z carries one such C_2 axis into another.

In summary, the following twenty symmetry operations are present in the structure of the molecule ferrocene: E, $C_5(= S_{10}^2)$, C_5^2, C_5^3, C_5^4, S_{10}, S_{10}^3, $i(S_{10}^5)$, S_{10}^7, S_{10}^9, five σ_d's, and five C_2's. Taken together, these constitute the symmetry operations of the point group labeled D_{5d}.* By appropriate combination of these operations, it is easy to construct a multiplication table and prove to yourself that the operations listed for D_{5d} constitute a mathematical group. Table 1-2 is the multiplication table for the point group C_{2v}. Here each of the four operations leads to E when combined with itself. Thus E appears on the diagonal. In constructing multiplication tables, it is assumed that the product operation appearing at the intersection of any column with any row transforms an

Table 1-2 *Multiplication Table for* C_{2v}

Group C_{2v}	E	C_2	$\sigma_v(xz)$	$\sigma_v'(yz)$
E	E	C_2	σ_v	σ_v'
C_2	C_2	E	σ_v'	σ_v
$\sigma_v(xz)$	σ_v	σ_v'	E	C_2
$\sigma_v'(yz)$	σ_v'	σ_v	C_2	E

* In point group tables or *character* tables, the elements are grouped according to the classes to which they belong.

object in exactly the same way that the object would be transformed by sequentially performing the operations given at the top of the column and the left-hand side of the row. In the C_{2v} point group all operations *commute* with each other, for example, $\sigma_v(xz)C_2 = C_2\sigma_v(xz)$, and such a group is called Abelian. However, most groups of interest to us do not have this property.

POINT GROUP SYMMETRY

The point group to which a particular molecular structure belongs now can be determined in a systematic manner. The line chart Figure 1-14 indicates the procedure we will follow. First look for special types of symmetry such as are characterized by linear, cubic, and icosahedral structures. The linear molecule which contains a center of symmetry belongs to the point group $D_{\infty h}$; otherwise it is in $C_{\infty v}$. The infinity symbol implies that an infinite number of vertical planes may contain the molecule. Since the cubic groups T, O, T_d (tetrahedral), O_h (octahedral), and T_h require the presence of four tetrahedrally oriented three-fold axes, this symmetry is readily recognized, and so is the regular icosahedron, I_h, which has six five-fold axes.

Molecules which do not display any type of rotation axes belong to C_s, C_i, or C_1. In the latter case there is no symmetry at all; for the others only one element of symmetry is present in addition to the identity element. These point groups are groups of *order* two; that is, they have only two elements.

If the only symmetry operations present in addition to an n-fold rotation axis are $2n$-fold improper rotations, the point group is S_{2n}.

The presence of n two-fold rotation axes perpendicular to the *principal axis** places the molecule in a D point group—either D_{nh}, D_{nd}, or D_n. The subscripts indicate the symmetry planes, if any, that are present. Without the two-fold axes, the molecule belongs to C_{nh}, C_{nv}, or C_n, and again the subscripts indicate the presence or absence of symmetry planes.

The Schoenflies notation for a point group specifies the minimum symmetry required to define a particular point group. This

---

* Sometimes the principal axis is not unique, as for example in D_2 or T. In such cases the choice of axis to be called principal is arbitrary.

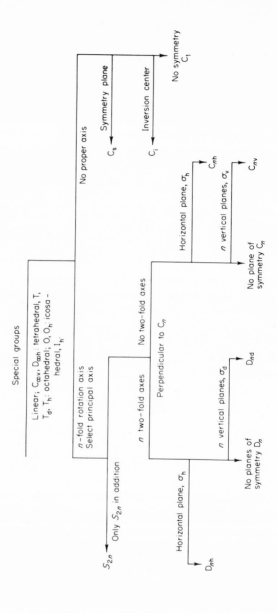

Figure 1-14 *Line chart identification of point group symmetry.*

fact is recognized quickly by the use of a mnemonic called a *stereographic projection.* (Stereographic projections originally were developed to describe crystal faces, but these details need not concern us here.) We will use stereographic projections simply to represent the symmetry operations present in point groups.

As an example of stereographic projection construction we consider the point group D_{4h} (Figure 1-15). The symmetry present

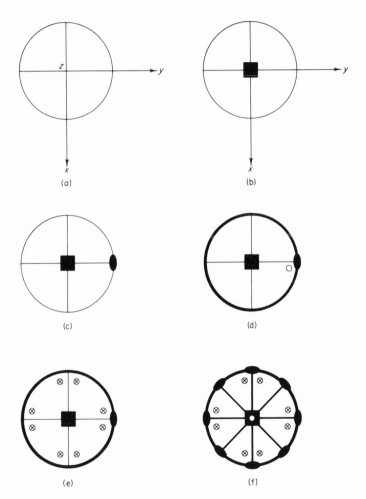

Figure 1-15 *Construction of the stereographic projection for the symmetry operations found in D_{4h}. (Reproduced by permission of the International Union of Crystallography.)*

will be indicated by symbols on a circle (Figure 1-15a). The z axis projects from the center of the circle. It contains the four-fold rotation axis and is given the symbol of a solid square, the appropriate regular polygon symmetric to rotation by $2\pi/4$ (Figure 1-15b). A two-fold axis is specified by a solid ellipse, a three-fold axis by a solid triangle, etc.

The point group D_{4h} requires the presence of a two-fold axis (Figure 1-15c) and a mirror plane perpendicular to the four-fold axis. The mirror plane is indicated (Figure 1-15d) by the heavy circle. Now, by placement of a general point O (x,y,z) as in Figure 1-15d, the three symmetry elements already listed cause the generation of fifteen other points (Figure 1-15e). Eight of the sixteen points lie above the plane of the paper and are labeled O, while those below are given the symbol ✕. It is now easy to add the other symmetry operations suggested by the arrangement of O's and ✕'s. Heavy lines represent planes, and the hole in the polygon indicates a center of symmetry. The number of points in the projection is equal to the number of symmetry operations present in the point group—its *order*.

An additional symbol is used when a rotation–reflection axis and a rotation axis are collinear but of different order. The rotation–reflection axis is labeled with an "empty" polygon, while the rotation axis is represented by the solid figure. Thus

indicates collinear S_6 and C_3 axes. Cubic groups require the introduction of additional curved lines that intersect at positions representing the corners of a cube inscribed within a sphere. The symbol for the point group T is presented in Figure 1-16.

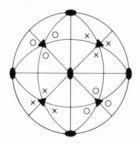

Figure 1-16   *Stereographic projection for the smallest cubic point group,* T. *(Reproduced by permission of the International Union of Crystallography.)*

Exercise 1-3   By means of a stereographic projection, develop the multiplication table for the point group C_{3v}.

Example   For C_{2v}, the stereographic projection is

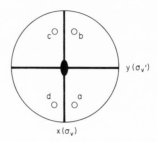

From this figure it is apparent that $C_2(\sigma_v')$ [point a] is σ_v'[point a] = point b, followed by C_2[point b] = point d. The same result is obtained by (σ_v)[point a] = point d. Thus, $C_2(\sigma_v') = \sigma_v$, etc.

Exercise 1-4   Deduce the point group for the molecular structures pictured in Figures 1-1, 1-2, 1-8, 1-9, and 1-12.

Exercise 1-5   The anion $NiBr_4^{2-}$ belongs to the point group T_d as a regular tetrahedron. Distortion of the anion by stretching one Ni—Br bond without bending the bonds and while leaving the other three bonds alone reduces the symmetry. Stretching two bonds simultaneously reduces the symmetry further. Considering all possible Ni—Br extensions, what point groups describe the structures? What point groups describe the structures of species in which Ni—Br bonds are allowed to be compressed relative to the equilibrium bond length?

Exercise 1-6   Carbon monoxide is linear, as is CO_2. What symmetry feature distinguishes these two molecules? Would this be true if CO_2 were C—O—O?

Exercise 1-7   A bunch of marbles on a tray can be made to pack together with all marbles in maximum contact. Maximum

density is achieved with a very symmetric structure.
What is the highest-order rotational axis found perpendicular to the tray? What others are present?
Can a five-fold axis appear in any packing arrangement
which has a well-defined number of marbles in a group
which repeats as a unit on the whole tray?

Exercise 1-8   By means of models, deduce the symmetry operations
present and the point group for the following molecular
species:

    (a) Tetrahedral $NiBrCl_3^{2-}$
    (b) Octahedral cis-$CoCl_3F_3^{3-}$
    (c) Octahedral $trans$-$CoCl_3F_3^{3-}$
    (d) Linear $HgCl_2$
    (e) Bent H_2S
    (f) Tetrahedral $CoCl_4^{2-}$
    (g) Planar cis-$PtCl_2Br_2^{2-}$
    (h) Planar $trans$-$PtCl_2Br_2^{2-}$
    (i) Trigonal bipyramidal PF_5
    (j) Trigonal bipyramidal cis-PF_3Cl_2
    (k) Trigonal bipyrimical $trans$-PF_3Cl_2
    (l) Square pyramidal $CuCl_5^{3-}$

Exercise 1-9   Sketch a stereographic projection for the point group
D_3. Add a center of symmetry to the figure. What new
points are developed? To what point group does the
new figure belong?

SUPPLEMENTARY READING

F. A. Cotton, "Chemical Applications of Group Theory." Wiley (Interscience), New
   York, 1963.
F. A. Cotton and G. Wilkinson, "Advanced Inorganic Chemistry," 2nd ed., pp. 116–122.
   Wiley (Interscience), New York, 1966.
F. C. Phillips, "An Introduction to Crystallography," 3rd ed. Wiley, New York, 1964.
G. H. Stout and L. H. Jensen, "X-Ray Structure Determination," Chap. 3. Macmillan,
   New York, 1968.

2 | *Atomic and electronic properties of d-block transition elements*

TRANSITION ELEMENTS

The thirty d-block transition elements that form the overwhelming majority of known coordination compounds all have one thing in common: They have energetically available "d-type" orbitals. These d orbitals participate in bond formation in the chemical compounds of the transition metals. Colors which are so distinctly characteristic of transition metal compounds appear as a direct result of the energies associated with these d orbitals. These d-block elements are called "transitional" in the sense that their electronic structure bridges the gap between the elements with filled s orbitals and those elements in which electrons occur in outermost p orbitals. The *first transition series* contains Sc, Ti, V, Cr, Mn, Fe, Co, Ni, Cu, and Zn.* Across this series the number of 3d electrons contained in the atom increases (Table 2-1). The *second transition series* begins with the element containing one 4d electron, yttrium (Y), and ends with the closed-shell structure of cadmium (Cd).

The *third transition series* begins with lanthanum (La), $[Xe]6s^2 5d^1$, and ends with mercury (Hg). However, the next element after La, cerium (Ce, No. 58), belongs to a series of elements which contains

---

* In some descriptions of the transition elements, Zn, Cd, and Hg are excluded since they contain no unfilled or partly filled d orbitals in their usual oxidation states. However, stereochemical features of compounds formed from these elements suggest pronounced d-orbital participation in the bonding. Hence they are included as members of the transitional element series in this book.

Table 2-1   *d-Block transition elements*

Atomic number	Element	Symbol	Configuration
21	Scandium	(Sc)	$1s^2 2s^2 2p^6 3s^2 3p^6 4s^2 3d^1$
22	Titanium	(Ti)	$[Ar]4s^2 3d^2$
23	Vanadium	(V)	$[Ar]4s^2 3d^3$
24	Chromium	(Cr)	$[Ar]4s 3d^5$
25	Manganese	(Mn)	$[Ar]4s^2 3d^5$
26	Iron	(Fe)	$[Ar]4s^2 3d^6$
27	Cobalt	(Co)	$[Ar]4s^2 3d^7$
28	Nickel	(Ni)	$[Ar]4s^2 3d^8$
29	Copper	(Cu)	$[Ar]4s^1 3d^{10}$
30	Zinc	(Zn)	$[Ar]4s^2 3d^{10}$
39	Yttrium	(Y)	$[Kr]5s^2 4d^1$
40	Zirconium	(Zr)	$[Kr]5s^2 4d^2$
41	Niobium	(Nb)	$[Kr]5s^1 4d^4$
42	Molybdenum	(Mo)	$[Kr]5s^1 4d^5$
43	Technetium	(Tc)	$[Kr]5s^2 4d^5$
44	Ruthenium	(Ru)	$[Kr]5s^1 4d^7$
45	Rhodium	(Rh)	$[Kr]5s^1 4d^8$
46	Palladium	(Pd)	$[Kr]4d^{10}$
47	Silver	(Ag)	$[Kr]4d^{10} 5s^1$
48	Cadmium	(Cd)	$[Kr]4d^{10} 5s^2$
57	Lanthanum	(La)	$[Xe]6s^2 5d^1$
72	Hafnium	(Hf)	$[Xe]4f^{14} 6s^2 5d^2$
73	Tantalum	(Ta)	$[Xe]4f^{14} 6s^2 5d^3$
74	Tungsten	(W)	$[Xe]4f^{14} 6s^2 5d^4$
75	Rhenium	(Re)	$[Xe]4f^{14} 6s^2 5d^5$
76	Osmium	(Os)	$[Xe]4f^{14} 6s^2 5d^6$
77	Iridium	(Ir)	$[Xe]4f^{14} 6s^2 5d^7$
78	Platinum	(Pt)	$[Xe]4f^{14} 6s^1 5d^9$
79	Gold	(Au)	$[Xe]4f^{14} 6s^1 5d^{10}$
80	Mercury	(Hg)	$[Xe]4f^{14} 6s^2 5d^{10}$

4f electrons. This series is called the rare earth or lanthanide series. Lanthanum itself displays chemical properties closely similar to the other elements in the lanthanide series. There is also the transition series called the actinide series, elements Nos. 90–103. Cations of the lanthanide and actinide series of elements readily form coordination compounds, but their chemistry is much less well developed than that of the d-block elements. Many of the symmetry and bonding concepts used with the d-block elements also may be appropriately applied to coordination compounds of f-block elements. However, there are some specific differences to be recognized which arise because f electrons participate less effectively in bonding than d electrons. The purpose of this chapter is to set the stage for applying symmetry concepts to the descriptions of the electronic structure and bonding in coordination compounds. No attempt has been made to treat f-block elements here, and the examples used later will be taken entirely from d-block transition elements.

PERIODIC BUILD-UP

As the atomic number increases in each of the three d-block transition series, the nd orbital is stabilized relative to the $(n + 1)$s orbital. The nd orbital is less stable than the $(n + 1)$s orbital at the beginning of a series, but becomes more stable at the end of the series. Progressing along a series of elements, say Ca to Zn, there is a regular increase in the number of nd electrons except when the nd and $(n + 1)$ orbitals have comparable energy, as at Cr or Cu (Table 2-1). These discontinuities occur at lower atomic numbers in the second and third transition series than in the first, due to the relatively greater stability of the 4d and 5d orbitals. The $[Kr]4d^{10}$ configuration in Pd, for example, is some 6564 cm^{-1} * more stable than the $[Kr]4d^95s^1$ configuration, and 25,100 cm^{-1} more stable than $[Kr]4d^85s^2$, the configuration which might have been expected from knowledge of the configuration of Ni (which is in the same *triad*, along with Pt).

When the nd orbitals are filled with ten electrons, as in Zn, Cd or Hg, they are sufficiently depressed in energy relative to $(n + 1)$s or $(n + 1)$p orbitals that to a first approximation they form part of a nearly "inert" electron core which has little influence on chemical properties. (In Ca, Sr, and Ba, the nd orbitals just begin to become energetically accessible, especially in the ions, and start to influence their chemistry.)

Just as the increased nuclear charge of the atom tends to lower the energy of the nd orbital relative to the $(n + 1)$s orbital, the overall positive electronic charge of a transition metal cation stabilizes the d orbitals relative to the $(n + 1)$s orbitals. This may produce an electron configuration for the *ground* or *lowest energy state* of the ion which is quite different from the configuration expected based on a systematic filling of the orbitals. For example, titanium, $[Ar]4s^23d^2$, as a 3+ cation might be

---

* One wave number, cm^{-1} (sometimes called a kayser, K), is equivalent in energy to 1/8067 eV or 1/349.5 kcal mole^{-1}.

expected to have the configuration [Ar]4s. However, its ground state configuration is in fact [Ar]3d; that is, the nd orbital is lower in energy than the $(n + 1)$s orbital. This increase in d-orbital stability with ionic charge influences the stereochemistry of many transition metal compounds.

ELECTRONIC PROPERTIES OF THE ATOMS AND IONS

The best approach available at present that gives a relatively useful understanding of the electronic properties of transition metal compounds starts with an assumption that the orbitals involved are identical in form to the orbitals calculated* for the simplest of atomic species, *hydrogen.*

The electronic properties of atomic hydrogen are completely determined by solving the quantum equation $\mathcal{H}\psi = \epsilon\psi$, where $\mathcal{H}$ (an operator called the Hamiltonian) is related to the kinetic and potential energy of an electron moving in the vicinity of a charged nucleus and the constant ϵ is the *quantized* energy that results. The ψ's, or wave functions, which solve this equation (named for its originator, Schrödinger) are described by their quantum numbers. The letter n gives the value of the principal or radial quantum number and has values from 1 to ∞; l, the azimuthal or angular quantum number, has values from 0, 1, 2, etc., to $n - 1$; m_l, the magnetic quantum number, has values ranging from 0, ±1, etc., to $\pm l$. No two electrons in the same atom can have the same n, l, and m_l unless they have different spins.

In Table 2-2 wave functions are given for the quantum numbers n, l, and m_l up to n = 3, l = 2, and $m_l = \pm2$. The position variables r, θ, ϕ, x, y, and z are defined by Figure 2-1; $R(r)$ describes the distance of the electron from the origin. This function changes for different values of n and l, but is independent of m_l. Thus the five 3d orbitals (n = 3, l = 2, $m_l = \pm2$, ±1, 0) all have the same value for $R(r)$ and are approximately equivalent in size or extension from the nucleus. The maximum value of ψ for each 4d orbital function occurs farther from the

---

* Very sophisticated quantum mechanical studies can now be done with high-speed digital computers. However, a complete electronic description of species containing as many electrons as are found in transition element compounds cannot be obtained at present without making some significant and often questionable assumptions. This condition is changing, so we may expect to obtain some new information to modify our models as computers continue to improve.

Table 2-2 *Angular form for hydrogenlike wave functions*

Quantum number				Angular wave function (made real and and normalized)	
n	l	m_l	Symbol	Polar	Cartesian
1	0	0	1s		$(4\pi)^{-1/2}$
2	0	0	2s		$(4\pi)^{-1/2}$
2	1	0	$2p_z$	$(3/4\pi)^{1/2}\cos\theta$	$(3/4\pi)^{1/2}(z/r)$
2	1	1	$2p_x$⎫	$(3/4\pi)^{1/2}\sin\theta\cos\phi$	$(3/4\pi)^{1/2}(x/r)$⎫ ^{a}
2	1	−1	$2p_y$⎭	$(3/4\pi)^{1/2}\sin\theta\sin\phi$	$(3/4\pi)^{1/2}(y/r)$⎭
3	0	0	3s		$(4\pi)^{-1/2}$
3	1	0	$3p_z$	$(3/4\pi)^{1/2}\cos\theta$	$(3/4\pi)^{1/2}(z/r)$
3	1	1	$3p_x$⎫	$(3/4\pi)^{1/2}\sin\theta\cos\phi$	$(3/4\pi)^{1/2}(x/r)$⎫ ^{a}
3	1	−1	$3p_y$⎭	$(3/4\pi)^{1/2}\sin\theta\sin\phi$	$(3/4\pi)^{1/2}(y/r)$⎭
3	2	0	$3d_{z^2}$	$(5/16\pi)^{1/2}(3\cos^2\theta - 1)$	$(5/16\pi)^{1/2}[(3z^2 - r^2)/r^2]$
3	2	1	$3d_{xz}$⎫	$(15/4\pi)^{1/2}\sin\theta\cos\theta\sin\phi$	$(15/4\pi)^{1/2}(xz/r^2)$⎫ ^{a}
3	2	−1	$3d_{yz}$⎭	$(15/4\pi)^{1/2}\sin\theta\cos\theta\sin\phi$	$(15/4\pi)^{1/2}(yz/r^2)$⎭
3	2	2	$3d_{x^2-y^2}$⎫	$(15/16\pi)^{1/2}\sin^2\theta\cos 2\phi$	$(15/16\pi)^{1/2}[(x^2 - y^2)/r^2]$⎫ ^{a}
3	2	−1	$3d_{xy}$⎭	$(15/16\pi)^{1/2}\sin^2\theta\sin 2\phi$	$(15/4\pi)^{1/2}(xy/r^2)$⎭

^{a} It is not strictly correct to assign m_l values to the real functions x, y, xz, yz, xy, and $x^2 - y^2$. For (n, l, m_l) values of (2, 1, 1) the angular function is $(3/4\pi)^{1/2}\sin\theta\, e^{i\phi}$. In linear combination with (2, 1, −1), $(3/4\pi)^{1/2}\sin\theta\, e^{-i\phi}$, the real functions given for $2p_x$ and $2p_y$ are produced. (When $a + ib$ is added to $a - ib$, $2a$, a real number results. Also, when $a - ib$ is subtracted from $a + ib$ and divided by i, the real number $2b$ is obtained.)

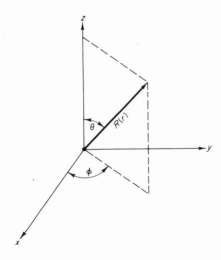

Figure 2-1 *Polar coordinates for hydrogenlike wave functions.*

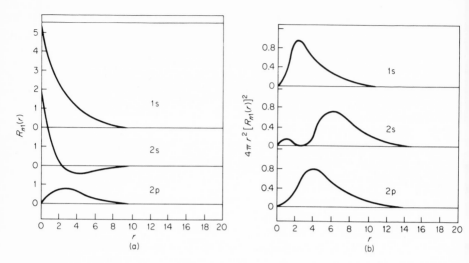

Figure 2-2   *Radial wave functions (a) and density distribution functions (b) for hydrogen-like* 1s, 2s, *and* 2p *functions (ordinates relative).*

nucleus than does the maximum value of each 3d function. However, both* $[R(r)]$ and $[R(r)]^2$ go to zero (a *node*) somewhere $0 < r < \infty$ for a 4d function. A 5d hydrogenlike function has two radial nodes between $r = \infty$ and 0, and the maximum value for $[R(r)]^2$ occurs even farther from the nucleus than for the 4d functions. Radial wave functions for hydrogenlike s and p functions and their corresponding distribution functions are given in Figure 2-2.

The angular dependencies of the hydrogenlike wave functions are important when considering the stereochemistry and bonding in transition metal compounds. For example, the 1s, 2s, and 3s functions, as seen in Table 2.2, are independent of the angles θ and ϕ. Since θ is a measure of the angle between $R(r)$ and the z axis, we see that the $2p_z$ function which is independent of ϕ is *isotropic* (same in all directions) in any plane perpendicular to z.

Exercise 2-1   Since $z = r \cos \theta$, it is apparent that $(\frac{3}{4}\pi)^{1/2} \cos \theta = (3/4\pi)^{1/2}(z/r)$. Thus the symmetry of the $2p_z$ function is identical to that of the z-directional vector. The

---

* Since ψ^2 is related to the probability that an electron exists, on the average, at a given position in space, its value is somewhat more meaningful than ψ in describing the extension of an orbital. Thus $[R(r)]^2$ has a simpler physical meaning than $R(r)$.

number r is independent of direction. In the d subshell one finds orbitals corresponding to the symmetries of xy, xz, and yz. The other three product functions, x^2, y^2, and z^2, sum to r^2 in a spherical geometry. Of the various linearly independent combinations one might make, the ones presented in Table 2-2 are very convenient. Show that

$$\left(\frac{15}{16\pi}\right)^{1/2}\left[\frac{z^2 - y^2}{r^2}\right] \quad \text{and} \quad \left(\frac{15}{16\pi}\right)^{1/2}\left[\frac{x^2 - z^2}{r^2}\right]$$

are equivalent to $3d_{x^2-y^2}$ and $3d_{z^2}$. Describe the shape of these functions.

Regions of space pictured to contain most of the electron density associated with a particular atomic orbital are given in Figure 2-3. These *conventional boundary surfaces* for s, p, and d functions are very useful in describing angular characteristics of hydrogenlike wave functions. Nodes in these hydrogenlike wave function (orbital) boundary surfaces are easily seen. Their presence can be determined mathematically by considering the angular portion of the wave functions as given in Table 2-2. If we examine the value of ψ^2 for the $3d_{xy}$ function, for example, we find that except when $\sin 2\phi = 0$, all θ values between 0 and π cause ψ^2 to be finite, while at $\theta = 0$ and π, ψ^2 is zero. There is a node along the z axis. Since $\sin 2\phi$ is zero when $\phi = 0$, $\pi/2$ (90°), π, or $3\pi/2$, there are also nodes at the xz and yz planes.

One additional symmetry feature of hydrogenlike wave functions should be recognized. Consider the algebraic sign associated with the wave function in any given region of space. *What happens to a wave function, for example, when the coordinates for a point* $[x, y, z]$ *are replaced by* $-x$, $-y$, *and* $-z$ $[\bar{x}, \bar{y}, \bar{z}]$? Here, ψ is unchanged for s and d functions, but becomes $-\psi$ for p and f functions. The former orbitals are said to be *symmetric with respect to inversion* through a symmetry center, while the latter are *antisymmetric to inversion*. This property of wave functions becomes important when we attempt to relate the probability that certain electronic transitions occur, to spectral intensities. It leads to the placement of plus and minus signs on lobes of figures such as Figure 2-3 that are meant to describe the probability density of the wave functions.

The maximum number of electrons that may be associated with any hydrogenlike atomic wave function is two. Consequently, we recognize that the symbol $1s^2$ corresponds to a filling of the 1s orbital

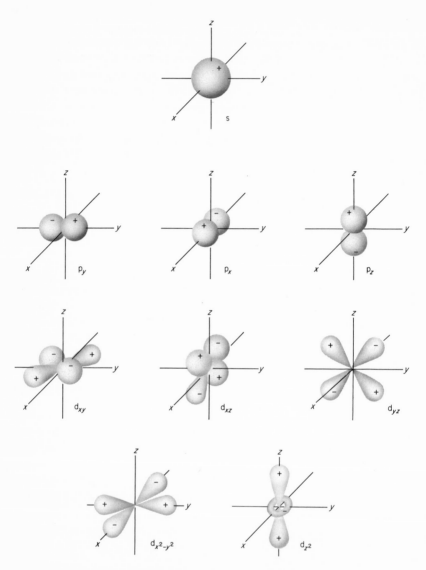

Figure 2-3 *Conventional boundary surfaces for hydrogenlike orbitals.*

by placement of two electrons into the atom with quantum numbers $n = 1$, $l = 0$, and $m_l = 0$. Similarly, associated with $n = 2$, $l = 1$, and $m_l = 0$, there are two possible electrons, $2p_z^2$. The quantum number $\frac{1}{2}$ is assigned to an electron to be proportional to its spin angular momentum.

In a magnetic field, there are two components to this angular momentum which are labeled $+\frac{1}{2}$ and $-\frac{1}{2}$. (The existence of only two components to the spin angular momentum for an electron was determined experimentally by O. Stern and W. Gerlach in 1921 when they showed that a beam of silver atoms, Ag:[Kr]$4d^{10}5s^1$, was divided into two equal beams by a magnetic field.) When an orbital is filled with electrons there is no resulting spin angular momentum as implied by the usage of $+\frac{1}{2}$ and $-\frac{1}{2}$ spin quantum numbers (m_s) for these electrons. With this description of the spin quantum numbers it can be said that the four quantum numbers associated with electrons on an atom must be different for each electron (the Pauli principle).

ENERGY CONSIDERATIONS

Solution of the Schrödinger equation for the hydrogen atom leads directly to quantized energy levels given by the expression

$$\epsilon = -\frac{1}{n^2}\left(\frac{2\pi^2 me^4}{h^2}\right)$$

where n is the principal quantum number; m is the mass of an electron,

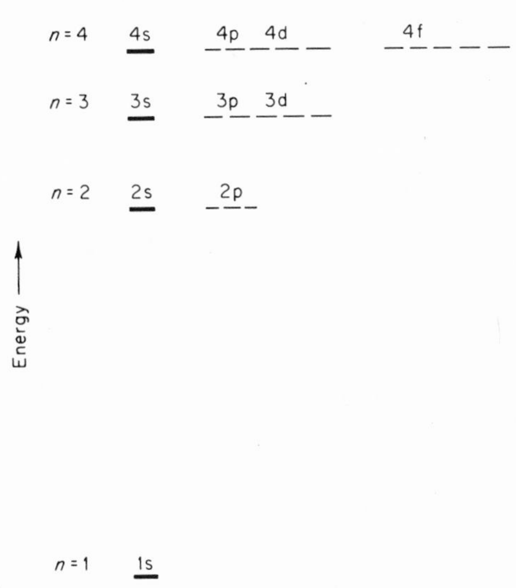

Figure 2-4 *Energy level diagram for the hydrogen atom.*

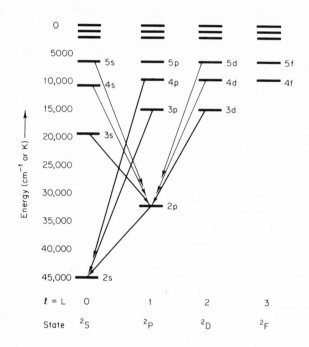

Figure 2-5   *Energy level diagram for lithium with strong transitions labeled.*

9.11×10^{-28} gm; h is Planck's constant, 6.626×10^{-27} erg-sec; and e is the electron charge, 4.8×10^{-10} esu. A set of energy levels for the hydrogen atom is given in Figure 2-4. Notice that these energy levels do not depend on l or m_l.

For a many-electron atom, the energy depends on other factors in addition to those dealt with in the two-particle hydrogen atom problem. For example, the nuclear charge is different from one and may even be nonspherical in its distribution. Also, the additional electrons exert coulombic repulsive forces on each other. Since electrons in s orbitals tend to occupy regions of space different from electrons in p or d orbitals, and p electrons different from d, etc., each electron type has a different influence on the *effective or average nuclear charge* felt by the other electrons. Electrons in different orbitals and/or different *shells** "shield" or "screen" the nucleus from other electrons by different amounts. Experimentally, it is observed that each shell and each orbital type within a shell (subshell)

---

* A shell corresponds to a particular value for n, the principal quantum number. Subshells correspond to ns, np, nd, etc., configurations within shells.

has a different energy. An energy level diagram such as the one for lithium (Figure 2-5) illustrates this fact.

ATOMIC SPECTRA

Soon after experimental studies began in the late 1800s, the electronic spectra of alkali metal vapors were recognized to resemble the spectrum for the hydrogen atom. From the expression for hydrogen atom energy levels, the energy difference, $\Delta\epsilon$, between any two levels, 1 and 2, is given by

$$\Delta\epsilon = R_{\mathrm{H}}\left(\frac{1}{n_1^2} - \frac{1}{n_2^2}\right) \tag{2-1}$$

where R_{H} is the Rydberg constant,* $2\pi^2\mu e^4/ch^3$ in cm^{-1}.

Alkali metal atoms have an outermost s shell containing only one electron; thus they resemble hydrogen. Their emission spectra consist of lines which decrease regularly in separation and intensity, and converge to a limit different from the limit for hydrogen. The series limit is the *ionization potential* of the atom. For hydrogen, this is

$$\frac{109{,}680 \quad cm^{-1}}{8067 \quad cm^{-1}eV^{-1}} = 13.6 \quad eV$$

The three most important series observed for lithium are the following (see Figure 2-5):

Sharp series: transitions from 2p, 3p, 4p, etc., to 2s (the ground state level)

Principal series: transitions from 3s, 4s, 5s, etc., to the 2p level

Diffuse series: transitions from 3d, 4d, 5d, etc., to 2p.

It was observed experimentally that the azimuthal quantum number, l, changes its value by only one unit (s → p, d → p, etc.) in each readily observed transition. This result produced a "selection

---

* Here μ, the reduced mass, $m_1m_2/(m_1 + m_2)$, has replaced m, and the constant includes the $(hc)^{-1}$ term needed to convert energy (in ergs) to cm^{-1}; h is Planck's constant, 6.63×10^{-27} erg-sec; and c is the speed of light in a vacuum, 3.00×10^{10} cm sec^{-1}. If the nuclear charge Z is different from one, $R_Z = (2\pi^2\mu e^4/h^3c)Z^2$.

rule" relating the intensity of a spectral transition to its "allowed" or "forbidden" character.

Exercise 2-2   Calculate the energy (in cm^{-1}) of the radiation emitted in the transition $3p \rightarrow 2s$ for He^+; for $3p \rightarrow 1s$ in Be^{3+}.

Hint   Since

$$\Delta\epsilon = R_H\left(\frac{1}{n_1^2} - \frac{1}{n_2^2}\right)$$

R_H must be calculated. For He^+, the reduced electron mass will be slightly larger than for H, and Z^2 will be four times as great. Thus $R_H \simeq 4 \times 13.6$ eV and $\Delta\epsilon \simeq$ $(5/36)$ (4×13.6) eV $\simeq 7.5$ eV $\simeq 60.5$ kK (kilo-kaysers).

Historically, it was found convenient to classify atomic states according to *terms* based on what is now recognized to be the value of L, a quantum number describing the *total orbital angular momentum* of the atom. For those atoms containing only one electron beyond a closed shell L equals l. These terms are labeled S, P, D, F, G, H, etc. depending on whether the angular momentum quantum number is 0, 1, 2, 3, etc. Hence the transition between the 3p level and the 2s level in lithium (Figure 2-5) would be labeled $^2P \rightarrow ^2S$. The number two in the upper left-hand corner of the symbol is designated as the multiplicity of the term, M. It is equal to one plus the number of unpaired electrons associated with the term $(n + 1)$ and describes the electron spin degeneracy. It is also directly related to the total spin, S, which may be found by multiplying the spin quantum number $\frac{1}{2}$ by the number of unpaired electrons; $M = 2S + 1$.

RUSSELL–SAUNDERS CLASSIFICATION

A commonly used classification scheme for atomic states of the transition element atoms and their ions is the Russell–Saunders (R–S) LSM_LM_S scheme, where M_L and M_S are defined as

$$M_L = m_{l1} + m_{l2} + m_{l3} + \cdots + m_{ln}$$

$$M_S = m_{s1} + m_{s2} + m_{s3} + \cdots + m_{sn}$$

(2-2)

This approach assumes that the spin angular momenta of individual electron spins interact negligibly with individual electron orbital angular momenta. In this scheme, m_{l1} is the value of the m_l quantum number for electron 1, m_{l2} is that for electron 2, etc. Now according to the solutions for the Schrödinger equation for hydrogen, m_l values for each electron depend on l itself and can vary from $+l$ to $-l$, including zero, by units of one. Since the total orbital angular momentum L arises from $2L + 1$ values of M_L from L, L $-$ 1, L $-$ 2, . . ., $-$L, each appearing at a different energy at a magnetic field, we speak of a state of L being $2L + 1$ times *orbitally degenerate*. Similarly, the total spin angular momentum S, for a given state can be considered as composed of $2S + 1$ vectorial components with values M_S of S, S $-$ 1, S $-$ 2, . . ., $-$S. The number of these components is the multiplicity of states that would be observed in a magnetic field. Thus a "doublet S state" having a term symbol ^{2}S (as observed for the lithium atom or the silver atom in their ground states) is orbitally nondegenerate, L = 0, but doubly spin degenerate, S = $\frac{1}{2}$, M = 2.

In determining the R–S terms for a given electron configuration, it must be remembered that *no two electrons can have the same four quantum numbers*. If they did, it would mean that both electrons have the same spatial properties with identical spins—an impossibility. Thus for two s type electrons, with n the same for both, a triplet S state, ^{3}S, is possible only if n is different for each electron, as for example $1s^2 2s^1 3s^1$. For p, d, and f type electrons, it is somewhat more tedious to determine the various possible states for two or more electrons, since with an $(np)^2$ configuration beyond the core,* for example, each of the two p electrons normally would be permitted to take on values of m_l of 1, 0, or -1, so that L may be as large as 2 in this case. The Pauli principle, however, demands that caution be exercised in writing down the possible terms, since if both electrons have the same m_l, they must have different spin.

Several schemes have been suggested for electron bookkeeping purposes; some of these may be deduced readily, but none will be described here.† Table 2-3 lists terms that may arise for various combinations of equivalent electrons.

---

* A core is a set of completely filled levels. As such it gives no angular momentum to the system. A filled shell of two s, six p, or ten d electrons produces a core. In a core there is a spherical symmetry of electron density distribution. Filled shells and subshells produce only isotropic influences on the electronic states and physical properties of the atoms.

† A student with access to a digital computer may readily generate all the terms possible for a given configuration, since the computer quickly scans all possible combinations for each M_L and M_S while remembering the Pauli principle.

Table 2-3   *Terms for equivalent electrons[a]*

s^2	^1S										
p^2	^1S	^1D	^3P								
p^3	^2P	^2D	^4S								
d^2	^1S	^1D	^1G	^3P	^3F						
d^3	^2P	$^2D(2)$	^2F	^2G	^2H	^4P	^4F				
d^4	$^1S(2)$	$^1D(2)$	^1F	$^1G(2)$	^1I	$^3P(2)$	^3D	$^3F(2)$	^3G	^3H	^5D
d^5	^2S	^2P	$^2D(3)$	$^2F(2)$	$^2G(2)$	^2H	^2I	^4P	^4D	^4F	^4G	^6S

[a] Terms for configurations p^{6-n} are the same as p^n; d^{10-n} the same as d^n.

Exercise 2-3   Determine the possible terms for the $(2p)^2$ configuration (see Table 2-3) by writing down all possible values of n, l, and m_l for each electron (microstates) that are consistent with the Pauli principle. A convenient way to indicate the spin values and the m_l values for $(m_l)_1 = 1$, $(m_l)_2 = 0$, $(m_s)_1 = +\frac{1}{2}$, $(m_s)_2 = -\frac{1}{2}$ is $(1^+, 0^-)$. For p^2, there are $6 \times 5/1 \times 2 = 15$ microstates.

The terms that arise as a result of d electrons are of principal interest in studies of transition metal complexes. One d electron gives a ^2D state $(l = L, m_l = M_L)$. With more d electrons, L may reach as high a value as six (Table 2-3). The ground state terms for important first-row transition metal ions are given in Table 2-4.

Table 2-4   *Ground state terms for some important first-row transition metal ions*

Ion	Configuration[a]	Ground term	Ion	Configuration[a]	Ground term
Ti^{3+}	$3d^1$	^2D	Fe^{2+}	$3d^6$	^5D
V^{2+}	$3d^3$	^4F	Fe^{3+}	$3d^5$	^6S
V^{3+}	$3d^2$	^3F	Co^{2+}	$3d^7$	^4F
Cr^{2+}	$3d^4$	^5D	Ni^{2+}	$3d^8$	^3F
Cr^{3+}	$3d^3$	^4F	Cu^{2+}	$3d^9$	^2D
Mn^{2+}	$3d^5$	^6S	Cu^+	$3d^{10}$	^1S
Mn^{3+}	$3d^4$	^5D			

[a] Beyond argon core.

HUND'S RULES

For a given electron configuration, say $1s^2 2p^2$, several spectroscopic terms are permitted, in this case ^1S, ^1D, and ^3P (see Table 2-3). What are the relative energies of these terms? Based on various experimental and theoretical arguments, rules have been formulated which govern the ground state to be expected for a given electron configuration.

(a) *For equivalent electron terms, those with the greatest multiplicity are at the lowest energy.* Thus the ^3P is at a lower energy than the ^1S and ^1D in the above example, and it is the ground state.

(b) For terms in (a) of the same multiplicity, *those of greatest L are lowest.* Hence the ^1D would be at a lower energy in the $1s^2 2p^2$ configuration than the ^1S state if there were no triplet term. However, these rules are strictly valid only for ground terms within the R–S coupling scheme. In cases where Hund's rules break down, the R–S scheme also breaks down.

Within a classification scheme which gives rise to various M_L terms, it is necessary to further consider substates which appear from a coupling of spin and orbital angular momenta of the atom or ion. This total angular momentum is designated J, and its value is appended to the term symbol as a right subscript, M_{LJ}. For first-row transition elements the energy differences between states of different J are generally small relative to energy differences between different M_L states. This is not true, however, for heavier elements such as the lanthanides or third-row transition elements where energy differences between states of different J may be several thousand reciprocal centimeters. In such cases the R–S coupling scheme fails, and more complicated treatments are required.

In the R–S scheme, J takes on values of L + S, L + S − 1, L + S − 2 to | L − S |. Thus, for the ^3P term in the example above, the three states $^3P_2$, $^3P_1$, and $^3P_0$ are formed. In a magnetic field, each of these states further produces $2J + 1$ magnetic states, each of slightly different energy. However, since we will not examine the origin of magnetic states in this text, we simply recognize their presence here without further discussion.

A third rule similar to the rules by Hund can be presented for the relative energies of states with equivalent M_L but different J (Figure 2-6).

(c) *As J increases, the energy of the substate increases for less than half-full subshells but decreases as J increases for subshells that are more*

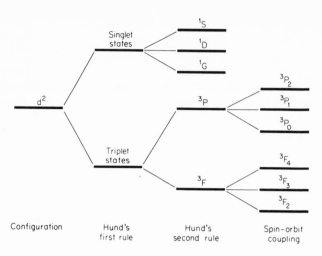

Figure 2-6 *Energies (not to scale) of Russell–Saunders terms for the equivalent electron configuration* $(nd)^2$. *Since Hund's rules strictly apply only to the ground state, the relative positions of the excited states may be interchanged for a given ion or atom.*

than half-full. Since the $2p^2$ configuration represents a subshell that is less than half-full, the energy progresses, $^3P_0 < \,^3P_1 < \,^3P_2$. For the configuration $1s^2 2p^4$, however, the energy is $^3P_2 < \,^3P_1 < \,^3P_0$ for the lowest-energy terms.

HOLE ELECTRON EQUIVALENCE

In Table 2-3 it is stated that configurations p^n and p^{6-n} or d^n and d^{10-n} produce identical terms. This fact can be understood if one recognizes that the absence of one electron in a shell or level is correctly described by assuming that a *positive hole* exists in the shell, with this hole having all the properties of the electron but a positive charge. Thus a "positron" d^2 configuration and an electron d^2 configuration produce states with identical angular momenta and terms. This symmetry about the half-filled shell greatly reduces the work associated with a description of electronic terms for many-electron atoms and ions.

Exercise 2-4   The Cu^{II} ion has an electron configuration $[core]3d^9$. What is the ground term for this ion?

MAGNETIC PROPERTIES

A detailed description of the magnetic properties of the transition elements and their alloys is outside the scope of this book. The emphasis here is placed only on the simplest magnetic effects observed for "magnetically dilute" transition metal ions in molecular species. This condition is met for many solutions of transition metal complexes and with crystalline solids which contain relatively few magnetic metal ions.

Four types of magnetic susceptibility behavior are observed with transition metal compounds (Table 2-5). We will consider only diamagnetism and paramagnetism, phenomena generally independent of the strength of the magnetic field. The other two types of magnetism, ferromagnetism and antiferromagnetism, require a concerted magnetic behavior by a large number of molecular units. Thermal agitation disrupts this process, producing a marked dependence on the temperature.

The magnetic susceptibility, χ, for a molecular system increases with the orbital and spin angular momenta. As a result, the greater the number of unpaired electrons, the larger the value of χ. For a mole of material (6.023×10^{23} molecules) the χ resulting is labeled χ_M.

The diamagnetic susceptibility of molecular species is found to depend almost entirely on the chemical constitution of the molecule. The negative of its value can be approximated quite accurately from a sum of additive constants for individual atoms and other characteristics of the molecule such as double bonds, triple bonds, etc. Some values of diamagnetic corrections are given in Table 2-6. By adding the diamagnetic correction to the experimental susceptibility for a species, the corrected molar susceptibility, χ_M^{corr} for the magnetic ion of interest is obtained. The χ_M^{corr} in turn is related directly to the corrected effective magnetic moment,

Table 2-5 *Magnetic behavior*

Type	Magnitude (cgs)	Field dependence
Diamagnetism	-1×10^{-6}	Independent
Paramagnetism	$1-100 \times 10^{-6}$	Independent
Ferromagnetism	$10^{-2}-10^{4}$	Dependent
Antiferromagnetism	$1-100 \times 10^{-6}$	Dependent

Table 2-6   *Diamagnetic corrections* $\times (10^6\ gm\text{-}atom^{-1})$

Na^+	6.8	Cl^-	23.4	$C{=}C$	-5.50
NH_4^+	13.3	Br^-	34.6	$C{\equiv}C$	-0.80
Ca^{2+}	10.4	NO_3^-	18.9	C (in benzene)	-0.24
Fe^{2+}	12.8	SO_4^{2-}	40.1	H	2.93
Co^{2+}	12.8	H_2O	13.0	C	6.00
Ni^{2+}	12.8	CO_3^{2-}	29.5	O (ketone)	-1.73
Cu^{2+}	12.8	OH^-	12.0	S	15.00

$\mu_{\text{eff}}^{\text{corr}}$, for the metal ion in the compound by

$$\mu_{\text{eff}}^{\text{corr}} = 2.8(\chi_M^{\text{corr}}T)^{1/2}\quad \text{B.M.} \tag{2-3}$$

where T is the temperature in degrees Kelvin and 2.8 is a constant which gives $\mu_{\text{eff}}^{\text{corr}}$ in units of Bohr magnetons (B.M.), $eh/4\pi mc$, or 0.927×10^{-20} erg G^{-1}. The magnetic moment is temperature-independent and directly related to the spin and orbital angular momenta of the ion by

$$\mu_{\text{L+S}} = [4S(S + 1) + L(L + 1)]^{1/2}\quad \text{B.M.} \tag{2-4}$$

Experimentally it is recognized that the effective orbital angular momentum associated with the metal ion for many transition metal complexes is nearly zero. As a result it is possible to describe the complexes in terms of the "spin-only" moments (2-5). For a system with one unpaired electron, μ_S becomes

$$\mu_S = \sqrt{4S(S + 1)}\quad \text{B.M.} \tag{2-5}$$

$\sim$1.73 B.M., a value close to the experimental $\mu_{\text{eff}}^{\text{corr}}$ for a great number of Cu^{II} compounds, for example. Since this value corresponds to $S = \frac{1}{2}$, the ground state is a spin doublet.

Exercise 2-5   The gram susceptibility of $Ni(O_2C_9H_{15})_2$ is $13.04 \pm 0.02 \times 10^{-6}$ cgs at 297°K. Calculate the effective magnetic moment for the complex assuming that the oxygen atoms are ketonic and that the carbon atoms are aliphatic.

Table 2-7 *Electron configuration and magnetic moment*

	d¹	d²	d³	d⁴	d⁵	d⁶	d⁷	d⁸	d⁹	d¹⁰
Ion ground state	²D	³F	⁴F	⁵D	⁶S	⁵D	⁴F	³F	²D	¹S
Number of unpaired electrons in *high-spin* octahedral complex	1	2	3	4	5	4	3	2	1	0
Spin-only moment (B.M.)	1.73	2.83	3.87	4.90	5.92	4.90	3.87	2.83	1.73	0
Number unpaired electrons in *low-spin* octahedral complex	1	2	3	2	1	0	1	2	1	0
Spin-only moment (B.M.)	1.73	2.83	3.87	2.83	1.73	0	1.73	2.83	1.73	0

Example For $NiCl_2$, the gram susceptibility at 293°K is 50.54×10^{-6} cgs. Multiplication by the formula weight, 129.6, gives an uncorrected molar susceptibility 6550×10^{-6}. The diamagnetic correction is $2 \times 23.4 \times 10^{-6}$ plus 12.8×10^{-6} or 59.6×10^{-6}. Thus χ_M^{corr} is 6610×10^{-6} and $\mu_{eff}^{corr} = 2.8(6610 \times 10^{-6} \times 293)^{1/2} = 3.9$ B.M. (This result is anomalous.)

Often, transition metal compounds produce a magnetic susceptibility close to the value calculated assuming that the metal ion is in the same atomic state as a "free" ion not bonded to other atoms. Such a system is labeled "spin-free" or "high-spin." Alternatively, some compounds indicate a reduction in the moment consistent with spin pairing and are called "spin-paired" or "low-spin" complexes. The reduction in the metal ion moment from high to low spin may be produced by atoms (ligands) bonded to the magnetic ion or by a magnetic coupling of two or more ions in the material. As a result the magnetic moment is meaningfully related to the chemical bonding in transition metal compounds. It will become clear later what this relationship is for mononuclear species since the stereochemistry of the compounds affects it. Table 2-7 anticipates these results for an octahedral arrangement of ligand atoms about the metal.

SUPPLEMENTARY READING

M. C. Day and J. Selbin, "Theoretical Inorganic Chemistry." Reinhold, New York, 1962.

G. Herzberg, "Atomic Spectra and Atomic Structure." Dover, New York, 1944.

3

Orbital symmetries
and their representation

ORBITAL SYMMETRY

In the preceding chapter electron configurations were presented for the ground state of d-block transition elements. These configurations assumed the presence of hydrogenlike atomic orbitals on the metal or ion. In transition metal coordination compounds the electronic structure of the central metal ion is described in terms of the properties of these same hydrogenlike orbitals. This assumption gives gratifying results so long as we are concerned primarily with symmetry properties of these orbitals. In this chapter we will learn to represent symmetry and particularly orbital symmetries in a manner consistent with the point group of a molecule.

While all the d orbitals with a given principal quantum number have identical energies in a "free" atom (see Chapter 2), this fivefold degeneracy is no longer required when the atom or ion is placed in an environment which is not spherically symmetric. In octahedral $CrCl_6^{3-}$, for example, each orbital associated with the chromium will be influenced to some degree by the presence of the ligands, but all five 3d orbitals need not be affected identically. The symmetry properties of these orbitals will conform to the geometry of the complex as a whole.

The degeneracy of the orbitals within levels having the same n and l quantum numbers is maintained only if each orbital in the

43

level has the same symmetry as every other orbital in the level. Thus, $2p_x$, $2p_y$, and $2p_z$ remain degenerate if they are transformed into each other by symmetry operations of the point group of the molecule. To state this another way, if orbitals become mixed up with each other when the various symmetry operations of the point group of the molecule are applied, the orbitals are symmetry related and remain degenerate (they cannot be separated) if they would have been degenerate in the free atom or ion. We can readily see that this is true for the three 2p orbitals in an octahedral complex. Consider a rotation about the C_3 axis coming out of the $+x$, $+y$, and $+z$ face of the octahedron (Figure 1-3). With x, y, and z coordinate axes as pictured, counterclockwise rotation by $2\pi/3$ (120°) carries p_x into p_y, p_y into p_z, and p_z into p_x. Repeating this rotation (applying C_3^2), p_x goes into p_z, p_y goes into p_x, and p_z goes into p_y.

It is important to notice that the sign of the 2p wave functions (see Table 2-2) does not change in the C_3 rotation mentioned above. Inversion, however, does change the sign of p (and f) orbitals but not that of the s or d orbitals. Test this by making $\theta = \pi + \theta$ for the $2p_z$ and the $3d_{z^2}$ angular wave functions. The s and d type of wave functions (or orbitals) are said to be even (*gerade*), labeled g, while the p and f type of functions are uneven (*ungerade*), labeled* u. The reader also should prove that the magnitude of a 3d function does not change when x is replaced by $-x$, y by $-y$, and z by $-z$.

Within the symmetry confine of an octahedral molecule, the d orbitals of a metal ion, unlike the p functions, lose some of their degeneracy. The set of functions labeled d_{xz}, d_{yz}, and d_{xy} (Figure 2-3) are carried into each other by a C_3 operation, but these functions never become part of the linearly independent set† formed from $x^2 - y^2$, $y^2 - z^2$, and $z^2 - x^2$, namely the functions (Table 2-3) labeled $d_{x^2-y^2}$ and $d_{z^2}(3z^2 - r^2)$. On the other hand, these last two orbitals become mixed up with each other under C_3. Try "mixing up" functions with some of the other operations for the O_h point group! It is easily seen, for example, that d_{xz} goes into d_{yz} upon performing a counterclockwise C_4 rotation about the z axis.

Exercise 3-1   Using polar coordinates, Figure 2-1, and the angular forms for the hydrogenlike wave functions (Table 2-2), show that in an octahedron, $C_3[2p_z] = 2p_y$.

---

* Odd functions have the quantum number $l = 1, 3, 5$; for even functions $l = 0, 2, 4$.
† See Exercise 2-1, Chapter 2.

Hint From Figure 2-1, $z = r \cos \theta$, $y = r \sin \theta \sin \phi$, and $x = r \sin \theta \cos \phi$. In a similar manner show that the octahedral C_4 rotation takes $3d_{x^2-y^2} \rightarrow -3d_{x^2-y^2}$.

As the symmetry present in the molecule decreases, the number of orbitals which can be "identically" degenerate,* as opposed to being accidentally equally energetic, decreases also. In fact, the point group to which a molecule belongs dictates the maximum degeneracy possible. For octahedral molecules this maximum degeneracy is three; for planar ones it is two; and for the icosahedral geometry it is five. Thus, in a square planar complex the three p orbitals cannot be degenerate. However, in tetrahedral or octahedral geometries, symmetry mixes them up. *Character tables* will help us determine the degeneracy patterns permitted for a given molecular symmetry.

REPRESENTATIONS OF SYMMETRY

It was mentioned in Chapter 1 that the elements associated with a particular point symmetry can be grouped into various classes. Each element in a class is related to other members of the class by a similarity transformation defined mathematically by $A^{-1}XA = Y$, where X and Y are in the same class and A is some element in the group. This similarity transformation amounts to a rearrangement of coordinates. The number of different classes that result dictates the number of primary ways in which we can represent a group. In fact, the number of these primary or irreducible representations is exactly equal to the number of classes in the group.†

---

* Orbitals interacting very little with the ligands may remain approximately degenerate even though symmetry requires the lifting of the degeneracy.

† There are several other rules of group theory which it will be useful to know and which can readily be verified using a character table. The mathematical proofs for these rules can indeed be very interesting, but they will not be presented here. As we find the rules to be useful, the various relationships will be introduced.

Example 3-1   Sketch the stereographic projection for the point group D_{4h}. Using this show that C_{2xy} and $C_{2\bar{x}y}$ are in the same class but in a different class from C_{2x} and C_{2y}.

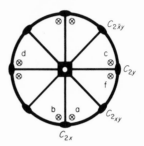

Consider point a^+; C_{2x} takes a^+ to b^-. The reciprocal of C_{2x} is C_{2x}^{-1} since $C_{2x}[\text{point } a^+] = [\text{point } b^-]$ and $C_{2x}^{-1}[\text{point } b^-] = [\text{point } a^+]$. Thus $C_{2x}^{-1}C_{2x} = (C_{2x})^2 = E$. Now, after $C_{2x}[\text{point } a^+] = [\text{point } b^-]$, carry out the operation C_{2xy}. This carries point b^- to point c^+. Following this by $C_{2x}^{-1} \equiv C_{2x}$, point d^- is obtained. Thus, $C_{2x}^{-1}C_{2xy}C_{2x}[\text{point } a^+] = [\text{point } d^-]$. As can be seen, the single operation $C_{2\bar{x}y}[\text{point } a^+]$ would have produced the same result. Thus, C_{2xy} and $C_{2\bar{x}y}$ are in the same class.

The second part of the problem is a little more complex. One possible but lengthy way to proceed would be to carry out all operations implied by $A^{-1}C_{2xy}A$ and show that none lead to C_{2x} or C_{2y}. While this procedure must lead to the correct answer, let us consider another approach. What is required if C_{2x} and, say, C_{2xy} are to be in the same class, $C_{2x} = A^{-1}C_{2xy}A$? From the stereographic projection this means that one carries out A on some point, prior to doing C_{2xy}, and then "undoes" A by doing A^{-1}. If C_{2xy} and C_{2x} are to be in the same class, they must be the same type of operation [for example, a counterclockwise rotation by $2\pi/4$ about some different related (by A) axis]. However, the axes of C_{2x} and C_{2xy} are *not* symmetry related in D_{4h}. (An eight-fold rotation axis would be required.) Thus, C_{2x} does not equal $A^{-1}C_{2xy}A$.

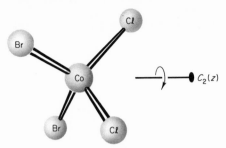

Figure 3-1 *The "tetrahedral" anion* $CoCl_2Br_2^{2-}$.

The "tetrahedrally" shaped molecule $CoCl_2Br_2^{2-}$ has a symmetry which places it in the point group C_{2v}. The symmetry operations which are present in C_{2v} are E, $C_2(z)$, $\sigma_v(xz)$, and $\sigma_v'(yz)$ (Figure 3-1). As each element is in a class by itself, there are four possible irreducible representations for this point group. One of these is easily deduced. Imagine, for example, a p_z type of orbital pointing along the z axis from the cobalt; we see that no operation of the group changes the function in any way. Thus, p_z must belong to a *totally symmetric* representation, that is, one that is invariant to each operation in the group. Such a representation is labeled A_1 (or A_g, or A_{1g}, depending on the particular point group) and must be present in every mathematical point group.

Table 3-1 is a character table* for C_{2v}; it represents the point group. By using the multiplication table (Table 1-2), it can be seen that each of the sets of numbers labeled A_1, A_2, B_1, and B_2 may represent the group.

Let us consider the last statement in more detail. By saying that A_1 represents the group C_{2v}, we mean that each of the numbers

---

* We can represent a group by matrices displaying the properties of the group. For example, the matrix $\begin{pmatrix} 1 & 0 \\ 0 & 1 \end{pmatrix}$ could represent the E operation if C_2, $\sigma_v(xz)$, and $\sigma_v'(yz)$ are also 2×2 matrices, since $\begin{pmatrix} 1 & 0 \\ 0 & 1 \end{pmatrix}\begin{pmatrix} A & B \\ C & D \end{pmatrix} = \begin{pmatrix} A & B \\ C & D \end{pmatrix}$. A character table contains the "traces" of similar square matrices. These traces are the sums of the elements along the diagonal of the matrices.

Traces of unitary matrices also display the properties of the group. The matrices required are unitary since such matrices display the special property that their determinant has an absolute value of 1 and is a root of unity. This requirement exists because if the matrix A is to represent a symmetry operation, $A^p = E$, the identity matrix. The exponent p is the number of times we must carry out A before we return to the initial symmetry condition. Since the determinant of E is $|E| = 1$, $|A^p| = |A|^p = 1$ and $|A| = (1)^{1/p}$.

Table 3-1   *Character table^a for the point group* C_{2v}

C_{2v}	E	C_2	$\sigma_v(xz)$	$\sigma_v'(yz)$		
A_1	1	1	1	1	z	x^2, y^2, z^2
A_2	1	1	-1	-1	R_z	xy
B_1	1	-1	1	-1	x, R_y	xz
B_2	1	-1	-1	1	y, R_x	yz

^{a} Symbols in the last two columns are explained later.

in the row of A_1 stands in the same relationship to every other number in the row as their respective symmetry operations. Thus the product $[C_2][\sigma_v(xy)]$ can be represented by 1×1 if $\sigma_v'(yz)$ also is represented by 1, since $[C_2][\sigma_v(xz)] = \sigma_v'(yz)$ (see Table 1-2). Similarly for B_2, $[C_2][C_2]$ is represented by $(-1)(-1)$ or 1; this product is E. Also, $[\sigma_v(xz)][C_2] = (-1)(-1) = (1) = \sigma_v'(yz)$.

The four representations required for the point group C_{2v} are the representations listed as A_1, A_2, B_1, and B_2. These representations are called irreducible since they cannot be reduced to the sum of still simpler representations of the group. The symbols used to designate irreducible representations are due to R. S. Mulliken and have the following meaning:

A and B are *one-dimensional** representations, E is *two-dimensional*, T is *three-dimensional*; A representations are *symmetric with respect to rotation about the principal C_n axis* and B representations are *antisymmetric* about this axis; the subscript g labels a representation as *symmetric with respect to inversion* and the subscript u labels it as *antisymmetric to inversion* (Table 3-2).

Another feature of character tables that is readily verified demands that the sum of the squares of the dimensions of the irreducible representations is equal to the number of elements (the order) in the group, $\sum\limits_{\text{irr. rep.}} d^2 = h$. The number of elements in the point group

C_{2v} is four. The O_h point group has 48 symmetry operations or elements. There are four one-dimensional, two two-dimensional, and four three-dimensional irreducible representations consistent with the above rule. The sum of the squares of these dimensionalities multiplied by the number of

* The dimensions correspond to the sizes of the square matrices from which the traces were obtained. The trace of the E unitary matrix then gives the dimension of the representation and the size of the matrix.

Table 3-2 *Symbols used with irreducible representations*

Dimension of representation	Character under			Symbol[a]
	E	C_n	i	
1	1	1		A
	1	-1		B
		1	1	A_u
		1	-1	A_g
2	2		1	E_g
			-1	E_u
3	3		1	T_g
			-1	T_u

[a] The symbol g originates with the German word *gerade*—symmetric to inversion, u originates with *ungerade*.

times each dimension appears, that is, $4(1)^2 + 2(2)^2 + 4(3)^2$, is 48. This relationship, along with the rule concerning the total possible number of irreducible representations in a group (that is, the number of classes) enables us to uniquely determine the dimensions of all the irreducible representations.

Exercise 3-2 List the twelve operations of the point group D_{3h} according to classes. How many one-, two-, and three-dimensional irreducible representations will appear in the character table for D_{3h}?

There are some additional properties of irreducible representations which are very important to a discussion of chemical bonding. The *orthonormal* nature of irreducible representations is one such property. The characters of the irreducible representations appear as if they are components of orthogonal and normalized vectors. That is, if each character for representation i is labeled $\chi_i(R)$, then the sum over all symmetry operations R of the product $\sum_R [\chi_i(R)][\chi_j(R)]$ is zero unless $i = j$, in which case $\sum_R [\chi_i(R)]^2 = h$, the number of operations (order) in the group. With C_{2v} (Table 3-1), $\sum_R [\chi_{A_1}(R)][\chi_{A_2}(R)] = 0$; that is,

$(1)(1) + (1)(1) + (-1)(1) + (-1)(1) = 0$. Also, we note that $\sum_{R} [\chi_A(R)]^2 = 4$, the order of the group.

These properties become important when we realize that orbitals in our transition metal compounds can be classified according to the irreducible representations present in the symmetry point group of the molecule.

As we stated before, the p_z orbital of the cobalt in $CoCl_2Br_2^{2-}$ is labeled A_1. It does everything the set of numbers in A_1 does under the operations of C_{2v}. Similarly, the p_y orbital can be represented by B_2 and the p_x by B_1. (The minus sign indicates that the function becomes the negative of itself upon carrying out the requisite operation.) Since these orbitals belong to orthogonal irreducible representations, they themselves must be orthogonal in the C_{2v} point group symmetry.

REDUCIBLE REPRESENTATIONS

It is important now to consider how we determine the irreducible representations spanned by a particular group of functions associated with a molecule. In the case above, it was seen that the three p orbitals, p_x, p_y, and p_z, span the irreducible representations B_1, B_2, and A_1 in C_{2v}. We could have arrived at this fact directly by constructing a *reducible* representation, T_r, for the set of three orbitals and reducing it into its irreducible components. By considering what happens to the set of three p orbitals together upon performing the operations of the group we will construct a representation which reduces to A_1, B_1, and B_2. This technique will enable us to produce proper reducible representations for any mathematical or physical property which can be depicted (mentally, verbally, or by pictorial representation), for example, chemical bonds, vibrations, orbitals, etc.

We learned long ago to use pictures to represent the symmetry properties of things. A sphere, for example, represents something that is the same in all directions from its center. Similarly, arrows on the three coordinate axes may be used to represent the symmetry of the p_x, p_y, and p_z orbitals. Changing the sign of the coordinate on an arrow inverts it. Probability density boundary surface pictures such as those in Figure 2-3 may be used to represent the symmetry of hydrogenlike orbitals. Using either arrows pointing along x, y, and z or the signed (plus and minus)

Table 3-3   *Reducible representation for* p *orbitals in* C_{2v}

C_{2v}	E	$C_2(z)$	$\sigma_v(xz)$	$\sigma_v'(yz)$
Γ_r	3	-1	1	1

boundary surfaces, we recognize that the symmetry operations may be applied to our pictorial representations of orbitals. Notice that the operation E does not change our pictorial representation.

Since the three p orbitals are each carried into $+1$ times themselves, the sum, 3, is placed under E. We need a three-dimensional representation to describe our three variables. Rotating the set of p orbitals about the z axis by C_2 causes the p_z orbital to remain unaffected but carries p_x into $-p_x$, and p_y into $-p_y$. Using the number $+1$ to signify that the function is carried into itself, the number -1 to indicate that it is carried into its inverse, and 0 to mean replacement by something completely new, the three numbers associated with the transformation of the p orbitals by $C_2(z)$ sum to -1. Similarly, under $\sigma_v(xz)$ and $\sigma_v'(yz)$, we find $+1$. Two orbitals remain unchanged under these last two symmetry operations, but one becomes inverted, hence the sum* is $+1$ (Table 3-3).

* These are the diagonal terms of the unitary matrix for C_2 which gives

$$C_2 \begin{bmatrix} p_x \\ p_y \\ p_z \end{bmatrix} = \begin{bmatrix} -p_x \\ -p_y \\ +p_z \end{bmatrix}$$

Thus,

$$C_2 = \begin{bmatrix} -1 & 0 & 0 \\ 0 & -1 & 0 \\ 0 & 0 & 1 \end{bmatrix}$$

since by the rules of matrix multiplication

$$\begin{bmatrix} -1 & 0 & 0 \\ 0 & -1 & 0 \\ 0 & 0 & 1 \end{bmatrix}\begin{bmatrix} p_x \\ p_y \\ p_z \end{bmatrix} = \begin{bmatrix} -p_x \\ -p_y \\ p_z \end{bmatrix}$$

The sum of the numbers on the diagonal (the trace or spur) of the matrix representing $C_2(z)$ is -1, the number of interest. Only when a function is transformed into itself or some fraction of itself will a nonzero term appear on the diagonal. In general we write a rotation matrix for rotation about z by any angle θ

$$C_\theta = \begin{bmatrix} \cos\theta & \sin\theta & 0 \\ -\sin\theta & \cos\theta & 0 \\ 0 & 0 & 1 \end{bmatrix}$$

(*Continued on page 52.*)

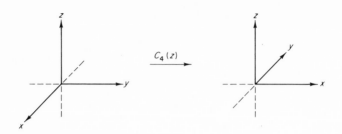

Figure 3-2   *Rotation of x, y, and z by $C_4(z)$. These coordinates represent p_x, p_y, and p_z.*

By deducing reducible representations in this way, we concern ourselves *only with the fraction of the function reproduced upon applying a particular symmetry operation.* The fact that certain functions in our set are carried completely into other members of the set simply reduces the labor involved. For example, a counterclockwise rotation of the three *p* functions (Figure 3-2) by $C_4(z)$ causes $p_x \rightarrow -p_y$ and $p_z \rightarrow p_z$. The number under C_4 for this representation will be 1 since p_x and p_y contribute nothing. For a C_3 rotation, however (Figure 3-3), $p_z \rightarrow p_z$, $p_x \rightarrow -\frac{1}{2}p_x + (\sqrt{3}/2)\, p_y$ and $p_y \rightarrow (-\sqrt{3}/2)\, p_x - \frac{1}{2}p_y$. The contribution

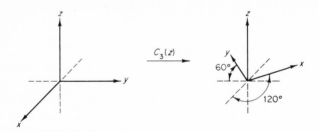

Figure 3-3   *Rotation of x, y, and z by $C_3(z)$.*

Hence rotation by $2\pi/2$ or C_2 leads to

$$\begin{bmatrix} -1 & 0 & 0 \\ 0 & -1 & 0 \\ 0 & 0 & 1 \end{bmatrix}$$

Rotation by C_3 equals

$$\begin{bmatrix} -1/2 & \sqrt{3}/2 & 0 \\ -\sqrt{3}/2 & -1/2 & 0 \\ 0 & 0 & 1 \end{bmatrix}$$

to the number under C_3 will be 1 for p_z and $-\frac{1}{2}$ each for p_x and p_y, giving a sum of 0. The fraction of p_z reproduced is 1, that of p_x is $-\frac{1}{2}$ and that of p_y is $-\frac{1}{2}$.

By comparing Table 3-1 with the representation for Γ_r in Table 3-3, we see by inspection that Γ_r can be formed as $A_1 + B_1 + B_2$ by summing up numbers in the same columns; Γ_r is said to be composed of the irreducible representations A_1, B_1, and B_2 in the point group C_{2v}. In this symmetry, the three p orbitals are not degenerate.

In general, whenever we are concerned with the number of independent ways in which we can represent a set of relationships which appear in a molecule, we will form a reducible representation for these relationships, consistent with the point group symmetry of the molecule, and reduce it. A direct procedure for reducing reducible representations now will be described.

Exercise 3-3   The N—H bonds in the ammonia molecule may be pictured by lines joining the atom centers. Deduce the reducible representations associated with these three bonds in the point group of the molecule.

REDUCTION OF REDUCIBLE REPRESENTATIONS

As we stated before, the five nd orbitals are not equivalent in an octahedral symmetry, but are split up into a group of size two, the e_g set, $(d_{z^2}, d_{x^2-y^2})$ and one of size three, the t_{2g} set (d_{xz}, d_{yz}, d_{xy}). These same d orbitals also are not equivalent in a square planar complex of D_{4h} symmetry. We are now in a position to construct a reducible representation for the d orbitals in D_{4h} symmetry and to determine a way in which these five orbitals can be represented by irreducible representations in this symmetry.

In the point group D_{4h}, there are sixteen symmetry operations. The character table (Table 3-4) lists these operations in classes, along with the ten irreducible representations of this point group. A figure representing a structure of D_{4h} symmetry is shown as Figure 3-4. The coordinate system and the various C_2 axes are labeled. The C_4 axis is coincident with the z coordinate. The σ_v symmetry planes include the x and y axes, while the σ_d planes bisect the angle between the x and y axes.

What happens to five hydrogenlike nd orbitals on M when we subject them to the operations of the D_{4h} point group? Assume

Table 3-4   *Character table for the point group* D_{4h}

D_{4h}	E	$2C_4$	C_2	$2C_2'$	$2C_2''$	i	$2S_4$	σ_h	$2\sigma_v$	$2\sigma_d$		
A_{1g}	1	1	1	1	1	1	1	1	1	1		$x^2 + y^2, z^2$
A_{2g}	1	1	1	-1	-1	1	1	1	-1	-1	R_z	
B_{1g}	1	-1	1	1	-1	1	-1	1	1	-1		$x^2 - y^2$
B_{2g}	1	-1	1	-1	1	1	-1	1	-1	1		xy
E_g	2	0	-2	0	0	2	0	-2	0	0	(R_x, R_y)	(xz, yz)
A_{1u}	1	1	1	1	1	-1	-1	-1	-1	-1		
A_{2u}	1	1	1	-1	-1	-1	-1	-1	1	1	z	
B_{1u}	1	-1	1	1	-1	-1	1	-1	-1	1		
B_{2u}	1	-1	1	-1	1	-1	1	-1	1	-1		
E_u	2	0	-2	0	0	-2	0	2	0	0	(x, y)	

that each of the d orbitals is labeled in the usual way (see Figure 2-3), so that the d_{z^2} orbital lies along the z axis, the d_{xy} orbital is in the molecular plane, etc. Since each of the d functions is symmetric to inversion, we can ignore this operation and all operations which are lost when the inversion operation is removed from the group. We are left with a smaller group, D_4 (Table 3-5), a subgroup, which is sufficient to give us the information we do not already know. In the D_{4h} "supergroup" we know the d orbitals must be found among the representations labeled A_{1g}, A_{2g}, B_{1g}, B_{2g}, and E_g because they are unchanged under inversion.

In the point group D_{4h} it is observed that the operations i, $2S_4$, σ_h, $2\sigma_v$, and $2\sigma_d$ are obtained directly by combining the operations E, $2C_4$, C_2, $2C_2'$, and $2C_2''$, respectively with i, the inversion opera-

Table 3-5   *Character table for* D_4

D_4	E	$2C_4$	$C_2(= C_4^2)$	$2C_2'$	$2C_2''$		
A_1	1	1	1	1	1		$x^2 + y^2, z^2$
A_2	1	1	1	-1	-1	z, R_z	
B_1	1	-1	1	1	-1		$x^2 - y^2$
B_2	1	-1	1	-1	1		xy
E	2	0	-2	0	0	$(x, y)(R_x, R_y)$	(xz, yz)
Γ_d	5	-1	1	1	1		
Γ_p	3	1	-1	-1	-1		

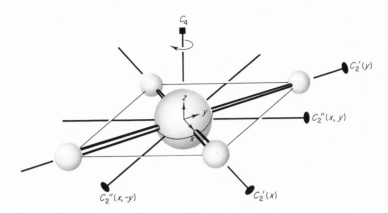

Figure 3-4 *Rotation axes in a square planar,* D_{4h} *symmetry structure.*

tion. This operation commutes with each operation of the subgroup D_4 as, of course, does E. The operations E and i constitute the point group C_i (Table 3-6). The supergroup D_{4h} can be formed directly from $D_4 \times C_i$ and is called the *"direct product* group" of D_4 and C_i. Table 3-4 is blocked out to illustrate this result. The upper left-hand corner is simply $D_4 \times (+1)$, the upper right-hand corner is $D_4 \times (+1)$, etc. In other words, the D_4 character table has been multiplied by the character of C_i in the same order as the symmetry operations have been combined. Direct product groups, $A \times B$, may be obtained whenever each operation in A commutes with every operation in B.

Stereographic projections are very useful in forming direct product groups. For example, the stereographic projection for D_4 (Figure 3-5), shows that there are eight symmetry operations. Adding an inversion center, each $\times$ becomes $\bigcirc$ and each $\bigcirc$ becomes $\times$ so that the full D_{4h} stereographic projection (Figure 1-15) with its sixteen operations is obtained.

Table 3-6 *Character table for* C_i

C_i	E	i		
A_g	1	1	R_x, R_y, R_z	$x^2, y^2, z^2, xy, xz, yz$
A_u	1	−1	x, y, z	

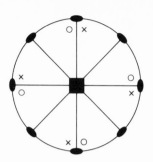

Figure 3-5   *Stereographic projection for the point group* D_4.

Exercise 3-4   Construct the stereographic projection for the point
group D_{3d} from the projection for D_3 by adding a symmetry center. Produce the D_{3d} character table.

Table 3-7 has been constructed to show how the d
orbitals transform under the operations of D_4. Note that only the orbitals
d_{xz} and d_{yz} are "mixed up" with each other. Since they are mixed up,
they are symmetry unseparable and belong to the two-dimensional representation E. The d_{z^2} orbital never changes (invariant), so it is in A_1, while
the $d_{x^2-y^2}$ and d_{xy} orbitals change sign under C_4. They each belong to a
different B representation since they behave differently under C_2' and C_2''.
The $d_{x^2-y^2}$ orbital belongs to B_1, while d_{xy} is B_2 in the character table.

If we now add the fact that the d orbitals are sym-
metric with respect to inversion (for example, Table 3-2) we have classified
the d orbitals according to the irreducible representations in D_{4h} as $A_{1g}(d_{z^2})$,
$B_{1g}(d_{x^2-y^2})$, $B_{2g}(d_{xy})$, and $E_g(d_{xz}, d_{yz})$. Instead of five degenerate orbitals,
as in the free atom, or sets of two and three degenerate orbitals, as in the
octahedral case, symmetry requires that all degeneracy except that of the
d_{xz} and d_{yz} functions be removed in D_{4h}. The irreducible representations
associated with x, y, z, xz, yz, xy, x^2, y^2, and z^2, and consequently orbitals
represented by these functions, are listed in the last two columns of the
character tables.

The above result has been obtained without even tabu-
lating a reducible representation for the five d orbitals in D_{4h}. With practice, however, such a representation, Γ_d, can be constructed and reduced
very rapidly without resort to a long table such as Table 3-7. To see how
this can be done, consider the nd orbitals in the symmetry group D_4.
Since there are five orbitals, the number under E for Γ_d is 5. Under the
operation C_4, $d_{z^2} \to d_{z^2}$, $d_{x^2-y^2} \to -d_{x^2-y^2}$, and $d_{xy} \to -d_{xy}$, but d_{xz} and

Table 3-7 *Transformation table for d functions in D_4*

D_4	E	C_4	C_4^3	C_2	$C_2'(x)$	$C_2'(y)$	$C_2''(x, y)$	$C_2''(x, -y)$
d_{z^2}	d_{z^2}	d_{z^2}	d_{z^2}	d_{z^2}	d_{z^2}	d_{z^2}	d_{z^2}	d_{z^2}
$d_{x^2-y^2}$	$d_{x^2-y^2}$	$-d_{x^2-y^2}$	$-d_{x^2-y^2}$	$d_{x^2-y^2}$	$d_{x^2-y^2}$	$d_{x^2-y^2}$	$-d_{x^2-y^2}$	$-d_{x^2-y^2}$
d_{xy}	d_{xy}	$-d_{xy}$	$-d_{xy}$	d_{xy}	$-d_{xy}$	$-d_{xy}$	d_{xy}	d_{xy}
d_{zz}	d_{zz}	$-d_{yz}$	d_{yz}	$-d_{zz}$	$-d_{zz}$	d_{zz}	$-d_{yz}$	d_{yz}
d_{yz}	d_{yz}	d_{zz}	$-d_{zz}$	$-d_{yz}$	d_{yz}	$-d_{yz}$	$-d_{zz}$	d_{zz}

d_{yz} go into each other. Using 0 and fractions of ± 1 to represent these observations, we have $+1$ for d_{z^2}, -1 each for $d_{x^2-y^2}$ and d_{xy}, and zero for the remaining two orbitals, for a sum of -1. Under C_2, d_{xz} and d_{yz} go into $-d_{xz}$ and $-d_{yz}$, respectively, while the other orbitals are unchanged; hence we have $+1$. Similarly, under C_2' and C_2'' we have $+1$. Thus the reducible representation Γ_d is given by the numbers 5, -1, $+1$, $+1$, and $+1$. For the three p functions the reducible representation, Γ_p, is 3, 1, -1, -1, and -1 (Table 3-5).

The mechanics of reducing or breaking up reducible representations into sums of irreducible representations of a point group follow directly from the observation that the irreducible representations themselves are orthogonal and normalized to h, the order of the group* as given by

$$\sum_R \Gamma_i^*(R)\,\Gamma_j(R) \;=\; \sum_R \chi_i^*(R)\chi_j(R) \;=\; \delta_{ij}h \tag{3-1}$$

$$(\delta_{ij} = 0,\; i \neq j;\; \delta_{ij} = 1,\; i = j)$$

By definition, then, the reducible representation, Γ_r, is a sum of irreducible representations. What is the result of the product of Γ_r with an irreducible representation Γ_i for the point group in question [Eq. (3-2)]?

$$\sum_R \Gamma_r(R)\,\Gamma_i(R) \;=\; \sum_R \chi_r(R)\chi_i(R) \tag{3-2}$$

If Γ_r is identical with or contains Γ_i, at least one term of the type given by Eq. (3-3) will appear. In fact, for every Γ_i contained in Γ_r there is a term equal to h. If Γ_i appears n times, then the sum is nh.

$$\sum_R [\chi_i(R)]^2 \;=\; h \tag{3-3}$$

In general, Γ_r will contain other irreducible representations in addition to Γ_i. These lead to products of the type given in Eq. (3-1) with $i \neq j$ and consequently contribute nothing to the numerical result. Thus, by taking the product of Γ_r with Γ_i and dividing by h, we

---

* Irreducible representations need not contain only real numbers, and indeed for some point groups, roots of unity such as $\pm i$, $e^{2\pi i/3}$, etc., appear in the character tables. In such cases Eq. (3-1) requires that one consider the product of a representation with its complex conjugate representation. Only then are the normalization and orthogonalization rules valid.

determine the number of times Γ_i appears in Γ_r [Eq. (3-4)]. By properly using this equation, reducible representations are readily reduced.

$$n_i = \frac{1}{h} \sum_R \chi_r(R)\chi_i(R) \qquad (3\text{-}4)$$

Consider the reducible representation, Γ_p, presented in Table 3-5 for the set of p orbitals in D_4 symmetry. By inspection we may recognize Γ_p to be formed from the sum of E with A_2 (adding characters under each operation). Let us use Eq. (3-4) to produce this result. First consider the product of Γ_p with A_1.

$$\sum_R \Gamma_p(R)\Gamma_{A_1}(R)$$

$$= 3(1) + [1(1) + 1(1)] - 1(1) + [-1(1) - 1(1)]$$

$$+ [-1(1) - 1(1)]$$

$$= 0$$

Since each element in a class has the same character, the coefficients at the top of the table can be used to give the same result:

$$\sum_R \Gamma_p(R)\Gamma_{A_1}(R)$$

$$= 3(1) + 2(1)(1) - 1(1) + 2(-1)(1) + 2(-1)(1) = 0$$

Thus, A_1 does not appear in Γ_p. Table 3-8 illustrates this manipulation

Table 3-8 The result of $\sum_R \Gamma_p(R)\Gamma_{A_1}(R)$ in D_4

D_4	①E	②C_4	①$C_2(= C_4^2)$	②C_2'	②C_2''	
Times Γ_p	×③	×①	×⊖①	×⊖①	×⊖①	
Times Γ_{A_1}	×①	×①	×①	×①	×①	
Equals $\sum_R \Gamma_p(R)\Gamma_{A_1}(R) =$	3	+2	−1	−2	−2	= 0

in another way. For A_2,

$$\sum_R \Gamma_p(R)\,\Gamma_{A_2}(R)$$

$$= 3(1) + 2(1)(1) - 1(1) + 2(-1)(-1) + 2(-1)(-1) = 8$$

Using (3-4), A_2 appears* once ($8/h = 1$) in Γ_p. Application of this equation to the irreducible representations B_1 and B_2 produces zero, and E gives one. Hence,

$$\Gamma_p = A_2 + E$$

Similarly, Γ_d (Table 3-4) reduces to

$$\Gamma_d = A_1 + B_1 + B_2 + E$$

It is common to label orbitals by the symbol of the irreducible representation (not capitalized) to which they belong. The d orbitals then would be given the symbols a_1, b_1, b_2, and e in the D_4 point

Table 3-9 *Orbital symmetries in some common point groups*

Point group	Atomic orbital	Symbol
O_h	s	a_{1g}
	p_x, p_y, p_z	t_{1u}
	$d_{x^2-y^2}$, d_{z^2}	e_g
	d_{xz}, d_{yz}, d_{xy}	t_{2g}
T_d	s	a_1
	p_x, p_y, p_z	t_2
	$d_{x^2-y^2}$, d_{z^2}	e
	d_{xz}, d_{yz}, d_{xy}	t_2
D_{4h}	s, d_{z^2}	a_{1g}
	p_x, p_y	e_u
	p_z	a_{2u}
	$d_{x^2-y^2}$	b_{1g}
	d_{xz}, d_{yz}	e_g
	d_{xy}	b_{2g}

* If the product results in a number not divisible by h to give a whole number, there is either an arithmetical mistake or the reducible representation is incorrect.

symmetry. For the point symmetry D_{4h}, the subscript g must be added to denote the fact that these d orbitals are symmetric under inversion as in a square planar complex. Table 3-9 lists the symbols for various atomic functions in some common symmetries. As stated before, this information is available in the last two columns in character tables such as those given in this chapter and in the Appendix.

Exercise 3-5   By constructing a reducible representation for the five d orbitals in T_d symmetry, show that they span the representations e and t_2.

Exercise 3-6   The positions of the three atoms in H_2O may be represented by nine vectors, three on each atom. Deduce the irreducible representations spanned by these nine position vectors in the C_{2v} symmetry of the molecule.

SUPPLEMENTARY READING

G. M. Barrow, "Molecular Spectroscopy." McGraw-Hill, New York, 1962.
F. A. Cotton, "Chemical Applications of Group Theory." Wiley (Interscience), New York, 1963.

Symmetry applied to molecular vibrations

4

Recognition of symmetry in a molecule causes us to expect certain specific results from a structurally dependent physical measurement. Conversely, the observation of results which fit a pattern suggested from symmetry considerations often leads us to make structural conclusions about the shape of a molecule. Infrared spectroscopy* is a physical technique that has contributed substantially to the understanding of molecular structure through the use of symmetry relationships. In an infrared spectrophotometer radiation is absorbed which vibrationally distorts the structure of a molecule without changing the distribution of electrons among its various electronic energy levels.

If we know the effect vibrational excitation has on the molecular symmetry we can begin to understand, for example, why HF absorbs infrared radiation strongly, while O_2 does not. Furthermore, we could assign the single absorption near 2000 cm^{-1} in tetrahedral $Ni(CO)_4$ to a transition involving primarily a CO bond "stretch." While we might expect to find five CO stretching vibrations in a molecule such as $Fe(CO)_5$ (Figure 4-1), symmetry readily leads us to deduce that only four are energetically discrete. Knowing that a vibration must produce a change in the dipole moment of the molecule before it will be excited by infrared radiation further enables us to understand (from symmetry relationships)

---

* The region of the electromagnetic spectrum categorized as "infrared" generally runs from 10,000 to 10 cm^{-1}. Some low-energy electronic transitions occur in the "near infrared," 10,000–4000 cm^{-1}, and some very energetic rotational and translational motions absorb energy independent of vibrational excitation in the region near 100 cm^{-1}.

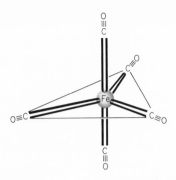

Figure 4-1 *The* D_{3h} *structure of iron pentacarbonyl,* Fe(CO)$_5$.

why only two of these four vibrations will be seen in the infrared spectral region.

REPRESENTING VIBRATIONAL SYMMETRY

The translational motion of an object in space can be described by means of a spacial vector which changes both its length and direction as the center of mass of the object changes. The same motion can be described equally well by noting changes that occur in the cartesian coordinates x, y, and z used to describe the position of the object in three-dimensional space. In general, it always will be necessary to ascribe three motion variables to translation unless, of course, we confine translation to a line or a plane where we would have one or two translational variables, respectively.

For a molecule we call the number of these position variables describing motion the "degrees of freedom" of the compound. In addition to three translational degrees of freedom, nonlinear molecules also have three rotational degrees of freedom. "Curved" arrows such as those pictured in Figure 4-2 may be used to represent the rotational variables R_x, R_y, and R_z for the molecule as a whole. For linear molecules such as CO_2 a rotation by any fraction of a degree, $2\pi/\infty$, about the C_∞ axis leads to a totally undistinguishable position. Consequently, there are only two rotational degrees of freedom for linear molecules which all belong to $D_{\infty h}$ or $C_{\infty v}$ point groups.

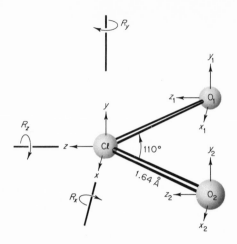

Figure 4-2   *Motional vectors on* ClO_2^-, *a species with* C_{2v} *symmetry.*

All motions of a molecule (translation, rotation, and vibration) may be described by noting separately the time-dependence of three position variables, x, y, and z, for each atom. In general there must be $3N$ position variables, where N is the number of atoms in the molecule. Since six of these variables correspond to motion of the molecule as a whole, $3N - 6$ ($3N - 5$ in the linear case) variables must be associated with internal motion in the molecule. In describing the internal motion of the molecule, we recognize that each atom is vibrating about its equlibrium position in space. The amplitude of this motion in general will be different for different atoms in the molecule. For some specific modes of vibration, however, the frequency and phase associated with each displacement coordinate will be the same. In other words, as one atom reaches the position of maximum displacement, the rest of the atoms in the molecule also are displaced a maximum amount from their equilibrium position. Also, all atoms pass through their equilibrium position at the same time. Such vibrations are called *normal* or *fundamental modes of vibration*. It is our purpose in this chapter to learn how symmetry may be used to describe these normal modes of vibration. Indeed, we shall learn that these modes correspond to irreducible representations in the point group symmetry of the equilibrium structure of the molecule.

We begin our attempt to categorize the internal motions of the atoms in a molecule by labeling three orthogonal position vectors on each atom. For the ClO_2^- anion pictured in Figure 4-2 there will be nine such vectors. We are free to position these vectors in any way

Table 4-1 *Representation of motion in C_{2v} symmetry for ClO_2^-*

C_{2v}	E	C_2	$\sigma_v(xz)$	$\sigma_v'(yz)$
Γ_{tot}	9	-1	1	3
Γ_{xyz}	3	-1	1	1
Γ_{rot}	3	-1	-1	-1
Γ_{vib}	3	1	1	3
Γ_{str}	2	0	0	2

we want. However, it will be convenient to consider a set in which all x_i's, y_i's, and z_i's are parallel.* Using the techniques described in Chapter 3, we readily produce the reducible representation, Γ_{tot} (Table 4-1) for this set of vectors. Reduction of this representation leads us to conclude that it contains three irreducible representations of A_1 type, one of A_2, two of B_1, and three of B_2. Thus, $\Gamma_{tot} = 3A_1 + A_2 + 2B_1 + 3B_2$. To consider only vibrations, we must now factor out those irreducible representations which characterize translational and rotational motion.

The reducible representations associated with molecular translation, Γ_{xyz}, and rotation, Γ_{rot}, are developed by choosing appropriate models which describe these motions. The arrows on Figure 4-2 do just this. The representations that result are given in Table 4-1. Since the vectors x, y, and z for any point group individually represent translation with x, y, or z directions, Γ_{xyz} is readily obtained from character tables containing columns listing the transformation properties of x, y, z, x^2, y^2, z^2, xz, yz, and xy. These functions are present in Table 3-1 (p. 48), the character table for the point group C_{2v}. Thus, $\Gamma_{xyz} = A_1 + B_1 + B_2$.

The rotational properties associated with a point group are listed as R_x, R_y, and R_z. From Table 3-1 it is seen that rotation about z, R_z, may be represented by A_2; rotation about x, R_x, by B_1; and rotation about y, R_y, by B_2. As it must, the sum $A_2 + B_1 + B_2 = \Gamma_{rot}$, the reducible representation for rotation listed in Table 4-1.

By removing those irreducible representations which relate to translational and rotational motion of ClO_2^- from Γ_{tot}, A_1, A_2, $2B_1$, $2B_2$, we are left with $\Gamma_{vib} = 2A_1 + B_2$. Consistent, then, with the $3N - 6$ rule, three vibrations ($N = 3$) are expected. Two of these vibrations are totally symmetric, A_1, and one is. antisymmetric (to the C_2 rotation), B_2. Thus, without further information about the mechanism

---

* Often it will be most convenient to choose one set of vectors to be directed along the interatomic axes.

by which radiation induces vibrational excitation, we do not expect vibrational degeneracy in ClO_2^-. We expect to see three bands.

Example 4-1   The molecule NH_3 belongs to the point group C_{3v}. Deduce the symmetry vibrational modes for the internal motions of the molecule.
The reducible representation for the $3 \times 4 = 12$ position vectors in C_{3v} is readily deduced.

C_{3v}	E	$2C_3$	$3\sigma v$
Γ_{tot}	12	0	2

Reduction produces $\Gamma_{tot} = 3A_1 + A_2 + 4E$. Since translation requires $A_1 + E$, and rotation requires $A_2 + E$, we are left with $\Gamma_{vib} = 2A_1 + 2E$. Thus there are $2a_1$ and $2e$ vibrations for NH_3.

Exercise 4-1   Deduce the symmetry vibrational modes for the metal—ligand bonds in the following molecules:

(a) tetrahedral $NiCl_4^{2-}$;

(b) tetrahedral $ANiCl_3^-$, where A is a ligand different from Cl;

(c) tetrahedral A_2NiCl_2, with A as in (b);

TYPES OF MOLECULAR VIBRATIONS

Since internal molecular motion involving major bond length changes ("stretches") generally require considerably more energy than bond angle changes ("bends") (the stretching frequencies will be higher than the bending frequencies), it is often desirable to reduce Γ_{vib} into a sum of irreducible representations representing molecular stretches, Γ_{str}, and bends. As we shall observe later, even though symmetry does not permit a strict mathematical separation of two vibrations belonging to the same irreducible representation, a large difference in frequency (or energy) between the two vibrational modes generally permits us to

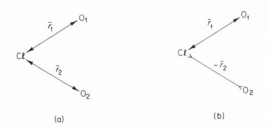

Figure 4-3 *Stretching of* Cl—O *bonds.*

consider each mode separately. Indeed, the two vibrations have little influence on each other even though symmetry permits their interaction.*

The reducible representation for the stretches in ClO_2^-, Γ_{str} (Table 4-1), is found by considering how the various operations in the point group C_{2v} change $\bar{r}_1$ and $\bar{r}_2$ (Figure 4-3a), lines which represent the Cl—O_1 and Cl—O_2 distances, respectively. The operations E and $\sigma'_v(yz)$ do not move $\bar{r}_1$ or $\bar{r}_2$, while C_2 and $\sigma_v(xy)$ interchange them.

The Γ_{str} produced reduces to $A_1 + B_2$. If we remove these symmetry representations from Γ_{vib}, we are left with a single representation, A_1, which must correspond to a bending motion of the molecule, since it cannot be a stretch.

In summary, group theory has enabled us to describe the internal vibrational motions of the anion ClO_2^- by the irreducible representations A_1 and B_2. The individual motions needed to produce these symmetry modes must be synchronized in frequency and phase; hence, they represent normal modes of vibrations. Since A_1 and B_2 are orthogonal representations, we recognize that vibrations of the type A_1 do not interact with those vibrations of B_2.

A stretching vibration with B_2 symmetry can be pictured by lengthening $\bar{r}_1$ and compressing $\bar{r}_2$ (see Figure 4-3b). Under the symmetry operations C_2 and $\sigma_v(xz)$ bond extension is interchanged with bond compression (the negative of itself). Upon performing E and $\sigma'_v(yz)$, the figure remains unchanged.

---

* The observation that two vibrations must have nearly identical frequencies to interact strongly is related to the example of the operatic singer shattering a good crystal glass. When the frequency of the singer's voice corresponds to a fundamental vibrational frequency of the glass, energy is transferred from the sound waves to the glass. However, those of us unable to achieve and maintain such a frequency with our voice fail miserably when we try this experiment.

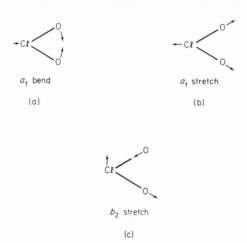

a_1 bend a_1 stretch

(a)   (b)

b_2 stretch

(c)

Figure 4-4 *Symmetry modes in* C_{2v} *for the vibrations of* ClO_2^-.

The a_1 stretch* of ClO_2^- has all of the symmetry of C_{2v} and can be represented by Figure 4-3a. The a_1 bend also must not be affected by the symmetry operations in C_{2v}. Arrows such as those pictured in Figure 4-4a describe such a motion. Since the molecule as a whole must not translate during this motion, a small arrow has been placed on Cl to represent its counteracting motion. The stretching motions are also pictured clearly by the arrows in Figure 4-4.

Exercise 4-2   Sketch the symmetry vibrational modes of NH_3. If we consider only N—H stretching motion, Γ_{str}, we see that this reduces to $A_1 + E$. Thus we can represent the internal motions in NH_3 by a stretch and a bend of A_1 symmetry and a stretch and a bend of E symmetry:

C_{3v}	E	$2C_3$	$3\sigma_v$
Γ_{str}	3	0	1

* A lower case letter corresponding to the capital letter of the irreducible representation of a vibration customarily is used to label the motion.

a_1 stretch

a_1 bend

e stretch

e bend

NORMAL VIBRATIONAL MODES

For a diatomic molecule the single vibration corresponding to stretching and compression of the chemical bond can be approximated (see Figure 4-5) by a harmonic oscillator potential energy, V, of the form

$$V = \tfrac{1}{2}kq^2 \tag{4-1}$$

where k is the restoring force and q is the displacement coordinate measured from the position of equilibrium. This motion in one dimension corresponds to the same motion displayed by two balls of mass m_1 and m_2 separated by a spring. The kinetic energy, T, of such a system is given by

$$T = \tfrac{1}{2}\mu\dot{q}^2 = \frac{1}{2}\left(\frac{m_1m_2}{m_1 + m_2}\right)\dot{q}^2, \qquad \dot{q} = \frac{dq}{dt} = \frac{d(r - r_e)}{dt} \tag{4-2}$$

where μ is $(m_1m_2/(m_1 + m_2))$, the reduced mass, and $\dot{q}$ is the rate of change with time of the coordinate.

If the masses are considered to be those of the individual atoms, quantum mechanics is applicable and the Schrödinger wave function, $H\psi = E\psi$, becomes

$$\frac{d^2\psi}{dq^2} + \frac{8\pi^2\mu}{h^2}(E - \tfrac{1}{2}kq^2)\psi = 0 \tag{4-3}$$

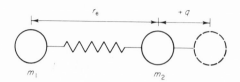

Figure 4-5   *Vibration of a simple system representing a diatomic molecule.*

Solution of this differential equation leads to the wave functions and energies given in Table 4.2.

A diagram describing the energy of the linear diatomic molecule as a function of internuclear distances is sketched in Figure 4-6. The minimum on this curve corresponds to the equilibrium distance of atom separation. Around this position the potential curve is quite well approximated by a parabola, the mathematical form of Eq. (4-1). For molecules in their lowest vibrational levels, the parabolic approximation usually leads to an acceptable description of the wave function. Since the energy difference between E_0 and E_1 for molecules of interest is approximately 300–4000 cm^{-1}, the thermal population at room temperature of

Table 4-2   *Wave functions and energies for the harmonic oscillator approximation to molecular vibration*

$$\psi_v = \frac{(\alpha/\pi)^{1/4}}{\sqrt{2^v v!}}\, e^{-\alpha q^2/2} Hv(\sqrt{\alpha} q)$$

$$E_v = h\nu_0(v + \tfrac{1}{2}) = \frac{hc}{\lambda_0}(v + \tfrac{1}{2})$$

$$E_0 = \tfrac{1}{2}h\nu_0 \qquad\qquad \psi_0 = \left(\frac{\alpha}{\pi}\right)^{1/4} e^{-\alpha q^2/2}$$

$$E_1 = \tfrac{3}{2}h\nu_0 \qquad\qquad \psi_1 = \left(\frac{\alpha}{\pi}\right)^{1/4} 2^{1/2} q e^{-\alpha q^2/2}$$

$$\alpha = \frac{2\pi\sqrt{\mu k}}{h} \qquad H_v(\sqrt{\alpha} q) \text{ is a Hermite polynomial of } v\text{th order } (v = 0, 1, 2, \ldots)$$

$$\nu_0 = \frac{1}{2\pi}\sqrt{\frac{k}{\mu}}$$

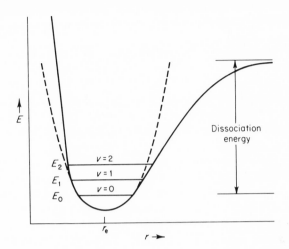

Figure 4-6 *Potential energy diagram for the ground state of a diatomic molecule. The dashed curve corresponds to the parabolic potential of simple harmonic motion.*

any levels other than E_0 is relatively small.* Consequently, we are concerned primarily with vibrational transitions from ψ_0 to ψ_1, $\Delta E = h\nu_0$.

Since molecules in their ground vibrational state display the symmetry of the equilibrium configuration, we need only examine the symmetry properties of ψ_1 to understand what changes occur in molecular structure under the $v = 0 \rightarrow v = 1$ vibrational excitation. Here, ψ_1 is linear in the coordinate, q, the coordinate used to describe the interatomic distance in the linear diatomic molecule. Consequently, for a diatomic molecule vibrating parallel to the x direction, the first excited vibration can be described by the symmetry properties of the vector x (q and x become identical).

For many-atom molecules it is appropriate to describe the motion in terms of mass weighted coordinates, $q_1 = \sqrt{m_1}\Delta x_1$, $q_2 = \sqrt{m_2}\Delta x_2$, etc. By suitable mathematical manipulation these coordinates can be transformed into a new set of coordinates based upon the normal vibrations of atoms present in the molecule. The wave functions and energies obtained are identical in form with those given in Table 4-2,

---

* From Boltzmann statistics, the ratio of excited molecules in $v = 1$ to those in the ground state, $v = 0$, assuming no degeneracy is, $\exp(-\Delta E/RT)$, where ΔE is the energy separation between the levels (in cal mole⁻¹), $R = 1.99$ cal mole-deg⁻¹, and T is the temperature (°K).

except q is now recognized to be a special coordinate describing the synchronous motion of several atoms in the molecule. These fundamental modes are constructed to be independent of each other. As in the diatomic molecule, the symmetry of the lowest vibrational wave function is that of the molecule, with each atom placed in its equilibrium position. The $v = 1$ excited vibrational level has a symmetry which corresponds to that of a normal coordinate. These normal coordinates can be represented by the orthogonal and normalized irreducible representations of the point group in question. The symmetry properties of these modes thus correspond to the symmetry properties of appropriate irreducible representations.

When we picture symmetry modes for vibration such as those for ClO_2^- in Figure 4-4, we are describing the symmetry properties of the lowest excited vibrational level. In other words, ClO_2^- shows a $v = 0$ to $v = 1$ vibrational transition corresponding to the symmetry change which occurs in going from the totally symmetric ground state structure to one of the structures pictured by Figures 4-4a–c. There are three such $v = 0$ to $v = 1$ transitions. While there is no change in symmetry in Figures 4-4a or b (they represent totally symmetric bending and stretching vibrations, respectively), in Figure 4-4c, the symmetry of the molecule is lowered from that of the ground state.

SELECTION RULES

We observed earlier that the three vibrations of ClO_2^- could be classified in C_{2v} symmetry as a_1 (stretch), a_1 (bend), and b_2 (stretch). Symmetry, and in particular the character table for C_{2v}, further enables us to understand why certain bands expected in the spectrum of a molecule either do not appear or appear very weakly. There is a "selection rule" associated with vibrational transitions. For infrared radiation, this selection rule says simply that in order for electromagnetic radiation to interact with a molecule and induce vibrational excitation the dipole moment of the molecule must be changed by the interaction. For a homonuclear diatomic molecule such as $Br_2(g)$, vibrational excitation clearly cannot change the dipole moment—there is no net dipole moment no matter how large we make the Br—Br distance. Such a molecule is "infrared inactive." For more complicated molecules, or diatomic molecules having a permanent dipole moment, infrared radiation will induce a net change in the dipole moment and be absorbed. From the symmetry properties of the normal modes of vibration, the possibility of dipole moment changes are readily deduced.

For the anion ClO_2^-, the a_1 (stretch) (Figure 4-4) produces a change in the dipole moment along the C_2 axes or z direction of the anion. The specific magnitude of the change in dipole moment will depend on the electronegativity of chlorine and oxygen atoms and on the changes in the Cl—O bond lengths that occur. While we cannot predict quantitatively the size of this effect we can be quite confident that the dipole moment will change, since Cl and O have different electron-attracting capabilities (electronegativity). The a_1 bend also should lead to a net dipole moment change, but its magnitude should be even smaller than for a_1 stretch. Consequently, the absorbance may be weak and difficult to find. The b_2 stretch, on the other hand, has a fairly large dipole moment change associated with it and should appear quite strongly. In $NaClO_2$ the a_1 stretch is found near 790 cm^{-1}, the a_1 bend near 400 cm^{-1}, and the b_2 stretch near 840 cm^{-1}.

INFRARED-ACTIVE VIBRATIONS

A more direct approach to infrared activity for vibrational transitions recognizes that the dipole moment of a molecule with a given symmetry must change under the symmetry operations in exactly the same way that x, y, and z change. In other words, $\Delta\bar{\mu}$, the change in the dipole moment, may be factored into $\Delta\bar{x}$, $\Delta\bar{y}$, and $\Delta\bar{z}$, its components in the x, y, and z directions. Since the equilibrium position may be called $x = y = z = 0$, the components of $\Delta\bar{\mu}$ have the same symmetry properties as x, y, and z.

As we discussed for the diatomic molecule, the $v = 0$ to $v = 1$ vibration corresponds to a linear change in the coordinate used to describe the motion of the molecule. In some cases the coordinate may be represented by x, y, and z. If so, the vibration requires a change in the dipole moment and will be *active* to infrared radiation.

$$\Gamma_{\text{vib}} = \Gamma_{x(\text{or } y \text{ or } z)} \tag{4-4}$$

Mathematically the requirement of (4-4) is met by such a vibration. In general, the mathematical requirement is that the product of the representations for the ground state, dipole moment, and the excited state,

Table 4-3   *Character table for the point group* D_{3h}

D_{3h}	E	$2C_3$	$3C_2$	σ_h	$2S_3$	$3\sigma_v$		
A_1'	1	1	1	1	1	1		$x^2 + y^2,\ z^2$
A_2'	1	1	-1	1	1	-1	R_z	
E'	2	-1	0	2	-1	0	(x, y)	$(x^2 - y^2,\ xy)$
A_1''	1	1	1	-1	-1	-1		
A_2''	1	1	-1	-1	-1	-	z	
E''	2	-1	0	-2	1	0	(R_x, R_y)	(xz, yz)

Γ_{prod}, (4-5) must contain the totally symmetric representation. If and only if this is true will the $v = 0$ to $v = 1$ transition be active.

$$\Gamma_{prod} = \Gamma v_{=1} \Gamma_{\bar{\mu}} \Gamma v_{=0} \tag{4-5}$$

To illustrate the power of this technique we will consider the CO stretching vibrations in $Fe(CO)_5$. The character table for D_{3h} is reproduced in Table 4-3. Following the usual rules, we generate the reducible representation given in Table 4-4 for the CO stretches. This reduces to $2A_1' + E' + A_2''$. Thus, we expect a maximum of four vibrations to appear near 2000 cm^{-1}, the CO stretching region, providing no transitions are inactive to dipolar radiation. However, inspection of Table 4-3 tells us immediately that no component of the dipole moment interacts with A_1'. Consequently, only the vibrations e' and a_2'' are active and observable* in the infrared.

Exercise 4-3   Deduce the infrared activity for the following vibrations:
  (a) all vibrations in planar $PtCl_4^{2-}$;
  (b) the CO stretches in $(\pi\text{-}C_6H_6)Cr(CO)_3$;
  (c) all vibrations in planar $cis\text{-}A_2PtCl_2$, A being a monodentate ligand different from Cl;
  (d) all vibrations in planar $trans\text{-}A_2PtCl_2$.

Table 4-4   *Reducible representation for CO stretches in* $Fe(CO)_5$

D_{3h}	E	$2C_3$	$3C_2$	σ_h	$3S_3$	$3\sigma_v$
Γ_{str}	5	2	1	3	0	3

---

* Since structural perturbations may occur in the solid or solution phases, it is sometimes possible to observe symmetry-forbidden infrared transitions in such samples.

STRUCTURAL CONCLUSIONS

The important lesson we learn from the above $Fe(CO)_5$ example is that the number of infrared-observable vibrations of modes is often considerably smaller than the total number of such vibrations that are possible. If $Fe(CO)_5$ had C_1 symmetry, none of the stretching vibrations would be forbidden. On the other hand, if $Fe(CO)_5$ were completely planar with D_{5h} symmetry, only one infrared-active transition should be observed. Consequently, the observation of two bands in the CO stretching region at 2028 and 1994 cm^{-1} is entirely consistent with the assumption of D_{3h} symmetry and not compatible with a possible D_{5h} formulation* (Figure 4-7).

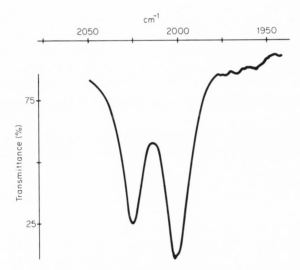

Figure 4-7   *The infrared spectrum of* $Fe(CO)_5$ *dissolved in* CCl_4 *in the CO stretching region. (Courtesy J. Levy and J. M. Burke.)*

* A square pyramidal C_{4v} structure predicts three infrared-active CO bands. Since the absence of a band could be caused by a failure of the technique to be sufficiently sensitive to a small dipole moment change, the C_{4v} structure could not be confidently ruled out by measurement of the infrared spectrum alone. This type of ambiguity generally is present when structures are deduced from spectroscopic measurements. F. A. Cotton has called such measurements "sporting techniques" for structure determination.

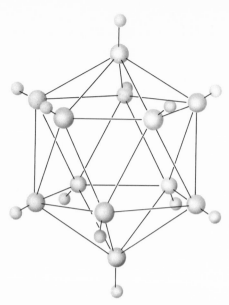

Figure 4-8   *Icosahedral* (I_h) *structure of the anion* $B_{12}H_{12}^{2-}$.

A structural consequence of symmetry is readily apparent. *The greater the molecular symmetry, the fewer the infrared-active vibrations.* This is one reason why tetrahedral CCl_4 is a good infrared solvent for solution studies. A particularly striking example of the effect of symmetry on the number of infrared-active bands can be seen in the infrared spectrum of $K_2B_{12}H_{12}$. The $B_{12}H_{12}^{2-}$ anion could show a maximum of 66 active vibrations. However, the very high symmetry, I_h (Figure 4-8), causes substantial vibrational degeneracy and the appearance of very few bands even in the solid state spectrum. M. F. Hawthorne has reported bands for $K_2B_{12}H_{12}$ at 2520 cm^{-1} (strong, B—H stretch), 1120 cm^{-1} (weak), 1075 cm^{-1} (strong), 755 cm^{-1} (weak), 748 cm^{-1} (weak), and 714 cm^{-1} (strong). Less than 10% of the $3N - 6$ bands possible with no symmetry. are observed. With the help of Figure 4-8 the reader may readily establish that there are four possible B—H stretching modes in I_h symmetry, a_g, h_g, t_{1u}, and t_{2u}, but only t_{1u} corresponds to a dipole moment change and consequently is infrared-active.

Exercise 4-4   The selection rule for Raman spectral intensity is that the excited vibration must belong to the same irreducible representation as x^2, y^2, z^2, xz, yz, xy, or combinations of these functions. Deduce the symmetry vibrational modes and indicate whether the vibration is active or inactive in the infrared and Raman for the following molecules:

(a) linear $ZnCl_2$;
(b) planar $FeCl_3$;
(c) pyramidal $FeCl_3$;
(d) dimeric $FeCl_3$.

SUPPLEMENTARY READING

F. A. Cotton, "Chemical Applications of Group Theory." Wiley (Interscience), New York, 1963.
K. Nakamoto, "Infrared Spectra of Inorganic Molecules." Wiley, New York, 1963.
E. B. Wilson, J. C. Decius, and P. C. Cross, "Molecular Vibrations." McGraw-Hill, New York, 1955.

Chemical bonding in transition metal compounds

5

As pointed out by Van Vleck some years ago, symmetry relationships are not subject to the usual uncertainties of chemical bonding theories. The physicist H. Bethe (in 1929) was the first to apply symmetry arguments to the "splitting of terms" and to formulate the purely electrostatic *crystal field theory*. In this theory it is assumed that the central metal ion is perturbed by an electrostatic interaction with point charges distributed around it so as to represent the ligands. Since the energetics of crystalline ionic compounds had been treated quite successfully by earlier workers assuming only electrostatic interactions between the ions, it was reasonable for Bethe to make the assumption that the metal–ligand interactions are largely coulombic. We now know that this approach inadequately describes metal–ligand bonding. Modifications must be made to allow for metal–ligand sharing of electrons (covalency). A crystal field theory which attempts to include covalency effects goes under the name *ligand field theory*. This theory was developed by chemists during the 1950s. Its main features come directly from the crystal field theory of the physicists.

CRYSTAL FIELD THEORY

Crystal field theory has been modified and developed substantially since 1929. It is presented here in an elementary form to help the neophyte student of coordination chemistry begin to develop a "chemical intuition" about electronic properties of transition metal compounds. In order to become an expert, the student must also build a substantial background in the fundamentals of atomic spectroscopy and quantum mechanics. As we have seen previously, if we surround a metal ion by ligand atoms to produce a complex having a specific geometry, some degeneracy originally present among the orbitals of the spherically symmetric metal ion is removed. For example, the five d orbitals are no longer equivalent in an octahedral complex but are split into t_{2g} and e_g orbitals containing three and two atomic orbitals, respectively. However, the magnitude of the splitting between t_{2g} and e_g may be very small. Symmetry tells us nothing about this. Furthermore, symmetry tells us nothing about the relative ordering of energy levels in the complex compared with the free ion. The crystal field theory model of chemical bonding attempts to answer some of these questions.

In principle, the Schrödinger equation can be formulated and solved assuming that the transition metal ion is surrounded by ligands represented by appropriate point charges. In addition to the potential energy terms present in the metal ion that are independent of the ligands, the Hamiltonian, $\mathcal{3C}$, requires an electrostatic potential energy term which averages the effect of all the charges associated with the ligands (Figure 5-1b). The charge is assumed to be distributed spherically about the metal ion. It causes a destabilization of all electronic states associated with the ion. However, the magnitude of this effect is difficult to determine

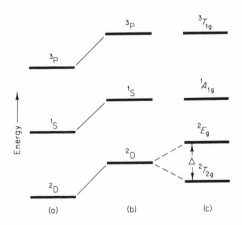

Figure 5-1 *Schematic representation of the energy of* (a) ^1S, ^3P, *and* ^2D *terms of an ion when subjected to electrostatic fields of* (b) *spherical and* (c) *octahedral symmetries.*

in actual practice. For an octahedral complex, the average destabilization may be several hundred kcal mole^{-1}.

The relative energies of ion terms may be changed by the spherical electrostatic field. These changes are similar in energy to the perturbations (10-100 kcal mole^{-1}) produced by the particular symmetry of the complex. It is this latter perturbation that contributes most to the description of the electronic spectra in the visible region and hence to the colors of transition metal compounds. Figure 5-1 illustrates the effect of these electrostatic fields on d-level terms.

Consider an octahedral complex of a transition metal ion which has one d electron beyond a spherically symmetric electron core. Its ground electronic state in the free ion will be ^{2}D according to the rules given in Chapter 2. In the complex, the five-fold orbital degeneracy of the ^{2}D state is removed. If this last electron is present in the e_g($d_{x^2-y^2}$ and d_{z^2}) orbitals of the ion, it will interact more strongly with the negatively charged ligands than if it were in the t_{2g}(d_{xz}, d_{yz}, and d_{xy}) orbitals. These latter hydrogenlike orbitals are not directed toward the ligands (Figure 2-3). Since coulombic energy is lost when electrons interact with negative charges, the e_g set of orbitals is destabilized relative to the t_{2g} set, (Figure 5-2).

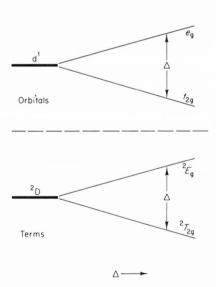

Figure 5-2 *Electronic states* ^{2}E$_g$ *and* ^{2}T$_{2g}$, *from the* [core]d^1 *configuration in an octahedral field.*

The placement of one electron in the t_{2g} orbitals produces a $^2T_{2g}$ state (capital "T"). This state is at somewhat lower energy than the ^2E_g state, arising from placement of the electron into e_g. The energy separating the states, Δ or 10Dq in the octahedral complex, is treated here as a phenomenological parameter. Its magnitude is dependent on properties associated with the metal ion and the particular ligand atoms which surround it.* The left superscript in a *symmetry term symbol* such as ^2E_g has the same meaning as it has for free ion terms, namely the spin multiplicity (see Chapter 2).

In octahedral fields, each free ion state produces certain specific symmetry states. The ion states important in transition metal chemistry produce the cubic symmetry states listed in Table 5.1. In general, the number of states arising from S, P, D, F, etc. is the same as the number of irreducible representations spanned by the s, p, d, f, etc., orbitals in the symmetry in question. In a cubic field one term each is given by the S and P ion states, while D states produce two new states. For the seven hydrogenlike f orbitals we find that the reducible representation with which they are associated in O_h symmetry reduces to a_{2u}, t_{1u}, and t_{2u}. Thus a ^2F state from a single f electron configuration produces $^2A_{2u}$, $^2T_{1u}$, and $^2T_{2u}$ states in an octahedral field. However, a ^2F state from a d electron configuration will produce $^2A_{2g}$, $^2T_{1g}$, and $^2T_{2g}$ states in O_h since the d orbitals are symmetric to inversion, not asymmetric like the f orbitals. Ignoring the symmetry associated with inversion (it can be added later), G states split into the terms A_1, E_1, T_1, and T_2 (Table 5-1).†

---

* Van Vleck and others have approximated Dq by the methods of quantum mechanics, but the quantitative results are generally only fair. See, for example, P. O. Offenhartz, *J. Am. Chem. Soc.* **91**, 5697 (1969).

† Rotation about the z axis causes a hydrogenlike wave function, $R(r)\theta(\theta)e^{im\phi}$, to change only in the $e^{im\phi}$ term. Upon rotation by an angle α, this term becomes $e^{im(\phi+\alpha)}$. For a given term with L total angular momentum (or l angular momentum for an individual electron) the range for M_L is $-L$, $-L + 1$, $-L + 2$, ..., $+L$ (m_l also goes from $-l$ to $+l$). Thus there are $2L + 1$ wave functions, each containing an $e^{iM_L\phi}$ term. The trace of the matrix which is associated with a rotation of these $2L + 1$ wave functions by α degrees (a) contains a geometric progression whose sum leads to (b).

$$\chi = e^{-iL\alpha}(1 + e^{i\alpha} + e^{2i\alpha} + \cdots + e^{i2L\alpha}) \tag{a}$$

$$\chi = \frac{\sin(L + \tfrac{1}{2})\alpha}{\sin(\alpha/2)} \tag{b}$$

This result requires that the reducible representation for an L = 2 (D state) have a -1 value for the C_3 operation (rotation by $\alpha = 2\pi/3$). Similarly, (b) gives a trace of 1 for the C_2 rotation operation and -1 for C_4 if L = 2. Since the choice of coordinates is arbitrary, the trace of the matrix for any proper rotation independent of axis direction is given by (b).

Table 5-1   *Reduction of free ion terms (states) in a cubic symmetry*

$S \rightarrow A_1$
$P \rightarrow T_1$
$D \rightarrow E + T_2$
$F \rightarrow A_2 + T_1 + T_2$
$G \rightarrow A_1 + E + T_1 + T_2$
$H \rightarrow E + 2T_1 + T_2$
$I \rightarrow A_1 + A_2 + E + T_1 + T_2$

As in atomic spectroscopy, we assume that the electronic states for metal ions in transition metal complexes arise from certain describable electron configurations. Thus $[\text{core}]t_{2g}$ gives a $^2T_{2g}$ state, $[\text{core}]e_g$ gives a ^2E_g state, etc., the configurations being determined by the filling of orbitals labeled according to the symmetry of the system. The symbols are those appropriate to the specific irreducible representations of the symmetry group. In a planar complex we may find a state labeled $^2B_{1g}$. It could arise from the configuration $[\text{core}]b_{1g}$, that is, one d electron

Table 5-2   *Terms for various cubic d^n electron configurations* [a]

$d^1(d^9)$	e	^2E
	t_2	$^2T_2$
$d^2(d^8)$	e^2	$^1A_1 + {}^3A_2 + {}^1E$
	et_2	$^1T_1 + {}^3T_1 + {}^1T_2 + {}^3T_2$
	$(t_2)^2$	$^1A_1 + {}^1E + {}^3T_1 + {}^1T_2$
$d^3(d^7)$	e^3	^2E
	e^2t_2	$2{}^2T_1 + {}^4T_1 + 2{}^2T_1$
	$e(t_2)^2$	$^2A_1 + {}^2A_2 + 2{}^2E + 2{}^2T_1 + {}^4T_1 + 2{}^2T_2 + {}^4T_2$
	$(t_2)^3$	$^4A_2 + {}^2E + {}^2T_1 + {}^2T_2$
$d^4(d^6)$	e^4	$^1A_1$
	e^3t_2	$^1T_1 + {}^3T_1 + {}^1T_2 + {}^3T_2$
	$e^2(t_2)^2$	$2{}^1A_1 + {}^1A_2 + {}^3A_2 + 3{}^1E + {}^3E + {}^1T_1 + 3{}^3T_1 + 3{}^1T_2 + 2{}^3T_2 + {}^5T_2$
	$e(t_2)^3$	$^1A_1 + {}^3A_1 + {}^1A_2 + {}^3A_2 + {}^1E + 2{}^3E + {}^5E + 2{}^1T_1 + 2{}^3T_1 + 2{}^1T_2 + 2{}^3T_2$
	$(t_2)^4$	$^1A_1 + {}^1E + {}^3T_1 + {}^1T_2$
d^5	t_2e^4	$^2T_2$
	$(t_2)^2e^3$	$^2A_1 + {}^2A_2 + 2{}^2E + 2{}^2T_1 + {}^4T_1 + 2{}^2T_2 + {}^4T_2$
	$(t_2)^3e^2$	$2{}^2A_1 + {}^4A_1 + {}^6A_1 + {}^2A_2 + {}^4A_2 + 3{}^2E + 2{}^4E + 4{}^2T_1 + {}^4T_1 + 4{}^2T_2 + {}^4T_2$
	$(t_2)^4e$	$^2A_1 + {}^2A_2 + 2{}^2E + 2{}^2T_1 + {}^4T_1 + 2{}^2T_2 + {}^4T_2$
	$(t_2)^5$	$^2T_2$

[a] The right subscript g is attached for octahedral symmetry.

is associated with $d_{x^2-y^2}$, a b_{1g} orbital in D_{4h}. However, as in atomic spectroscopy, placement of more than one electron into given orbitals may produce several terms. For octahedral and tetrahedral (cubic) complexes with d^1–d^5 configurations, these terms are reproduced in Table 5-2. The configurations d^9–d^6 produce results identical to those for d^1–d^4, respectively.

DIRECT PRODUCT REPRESENTATION

The irreducible representations associated with configurations such as $[core]t_{2g}^2$, $[core]e_g^2$, etc. are readily deduced from the character table for the symmetry of the system. Assume, for example, the electron configuration in O_h symmetry to be $[core]a_{1g}^2$. This means that two electrons are placed into a totally symmetric a_{1g} orbital on the metal. Since the a_{1g} orbital has no spatial degeneracy (it is like an S orbital), it can hold only two electrons which, according to the Pauli principle, must be spin-paired. The resultant electronic state must be a spin singlet. The term symbol reflects the spatial degeneracy of the configuration. Since there is no spatial degeneracy associated with a_{1g}^2, the term symbol must be ^1A or ^1B, for only the A and B type of irreducible representations reflect no spatial degeneracy (see Table 3-2).

The possible irreducible representations (spatial terms) associated with a given configuration are obtained by forming the direct product representation of the irreducible representations for the individual electrons in the configuration and reducing this representation (if it is not irreducible). For example, the e^2 configuration in the point group O implies the presence of two electrons each having a possible two-fold spatial degeneracy. (The total possible spatial degeneracy is $2 \times 2 = 4$.) By taking the direct product of E with itself, the reducible representation given in Table 5-3 is produced. The representation reduces to $A_1 + A_2 + E$. Since each electron is of e type, the Pauli principle prevents both singlets

Table 5-3 *Direct product representation*

O	E	$8C_3$	$6C_2$	$6C_4$	$3C_2(C_4^2)$
E	2	-1	0	0	2
$E \times E$	4	1	0	0	4

and triplets for each symmetry state. In fact, elementary statistics demands that the E state be a singlet with either A_1 or A_2 being triplet.*

DESCENT IN SYMMETRY

Spin multiplicities for crystal field terms may be deduced by a thought process (*gedanken* experiment) in which it is assumed that the symmetry of the complex can be reduced to the point where all spatial degeneracy is removed. By correlating the resultant low symmetry states with those of the geometry containing spatial degeneracy, a unique description of the spin and spatial degeneracy of the state is obtained.

Consider the e^2 configuration in T_d as our example. From a correlation table (Table 5-4) or a character table, we recognize the e orbital in T_d to become a_1 and b_1 orbitals in C_{4v} (a symmetry sufficiently low to remove all the spatial degeneracy of the A_1, A_2, and E terms of e^2). Now we may think of placing the two electrons into these orbitals as a_1^2, b_1^2, or $a_1^1 b_1^1$. From the Pauli principle, the states resulting from a_1^2 and b_1^2 configurations must be singlets, while either a singlet or a triplet is permitted with $a_1^1 b_1^1$. Utilizing direct product relationships, we see that $^1A_1$ (in C_{4v}) results for a_1^2 and b_1^2, while $^3B_1$ and $^1B_1$ states are permitted

Table 5-4 *Correlation table of O_h terms on reduction of symmetry*

O_h	T_d	D_{4h}	C_{4v}	D_3	C_{2v}
A_{1g}	A_1	A_{1g}	A_1	A_1	A_1
A_{2g}	A_2	B_{1g}	B_1	A_2	A_2
E_g	E	$A_{1g} + B_{1g}$	$A_1 + B_1$	E	$A_1 + A_2$
T_{1g}	T_1	$A_{2g} + E_g$	$A_2 + E$	$A_2 + E$	$A_2 + B_1 + B_2$
T_{2g}	T_2	$B_{2g} + E_g$	$B_2 + E$	$A_1 + E$	$A_1 + B_1 + B_2$

---

* The configuration e^2 is like having two boxes ☐☐ each capable of holding two electrons. There are four ways to put the first electron into these boxes ($+$ or $-$ spin in either box). The second electron can go into the boxes in any of three ways. The product (4×3) is twelve, but since the two electrons are indistinguishable, we must divide by 2. Thus there is a total multiplicity (space × spin) of six for the configuration. In general, the statistical combination leads to $n(n - 1)(n - 2)\cdots(n - m + 1)/1\cdot2\cdot3\cdots m$, states, where n is the number of possibilities for the first electron and m is the number of electrons. For t_2^3 the result is $6\cdot5\cdot4/1\cdot2\cdot3$, as implied by $^4A_2 + {}^2E + {}^2T_1 + {}^2T_2$ (Table 5-2).

for $a_1^1 b_1^1$. Since one of the $^1A_1$ states in C_{4v} must correlate with E in T_d, the latter state must be ^1E. Furthermore, the $^1B_1$ state (in C_{4v}) also correlates with ^1E (in T_d). Hence the triplet state in T_d is the state that correlates with $^3B_1$ (in C_{4v}). This is $^3A_2$. In summary, then, the states in T_d must be $^3A_2$, $^1A_1$, and ^1E.

GROUND STATES

While both [core]d^1 and [core]d^9 configurations produce the same cubic terms, ^2E and $^2T_2$, the ground state for each will be different. We can consider the [core]d^9 electron configuration to be identical to a [core]d^1 "hole" or positive electron description. We recognize, then, that coulombic energy is gained by placing this hole in the e_g orbital set as opposed to placement in t_{2g}. Thus the [core]t_{2g} hole configuration is energetically less favorable than the [core]e_g hole configuration and the ^2E state becomes the ground state as in octahedral [core]$3d^9$ Cu^{II} complexes (Table 5-2).

In the crystal field model, calculations may be made to determine the energies of the various electronic states arising from d-orbital configurations. For the ion [core]d^1 hole or [core]d^9 electron configuration in an octahedral symmetry, the two states $^2T_{2g}$ and ^2E_g move apart linearly with Δ (Figure 5-2). Since Δ generally is found to be near 10–20 kK for octahedral transition metal complexes, the electronic transition $^2T_{2g} \rightarrow ^2E_g$ corresponds to a visible (3800–7800 Å) absorbance.

A transition at 25 kK absorbs some of the blue region of white light and causes the complex to appear red. Copper(II) complexes, generally colored blue or blue-green, exhibit the $^2E_g \rightarrow ^2T_{2g}$ transition near 14,000 cm^{-1}. This transition absorbs the radiation on the red end of the visible spectrum, transmitting the blue. Yellow to red colored complexes absorb the high-energy visible radiation.

TANABE–SUGANO DIAGRAMS

For configurations arising from more than one d electron, the variation in energy of electronic states with Δ is not readily determined. Certain approximations must be made to describe the electron–electron interactions in the presence of the ligands. Some moderately successful semiempirical techniques have been described to perform these

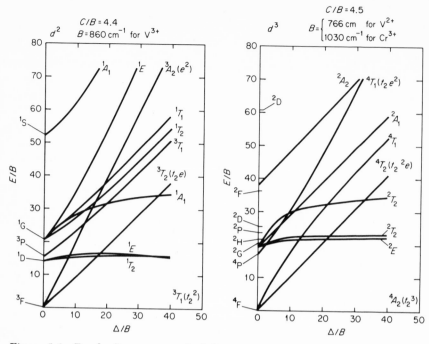

Figure 5-3 *Tanabe–Sugano energy level diagrams for the* d^2 *and* d^3 *configurations in cubic* (O_h) *symmetry.*

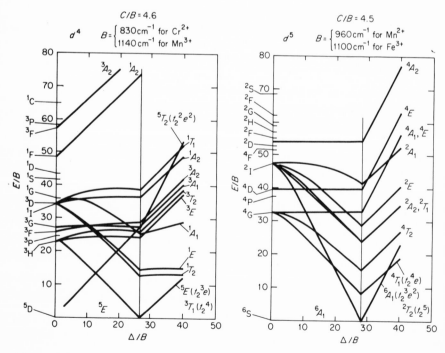

Figure 5-4 *Tanabe–Sugano energy level diagrams for the* d^4 *and* d^5 *configurations in cubic* (O_h) *symmetry.*

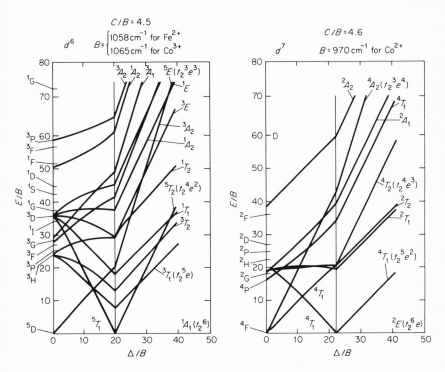

Figure 5-5 *Tanabe–Sugano energy level diagrams for* d^6 *and* d^7 *ions in a cubic* (O_h) *field.*

calculations. The results for the popularly used approach generally attributed to Tanabe and Sugano* are presented in the form of graphs (Figures 5-3 to 5-6). These diagrams for the ions d^2–d^8 in cubic fields are used quite successfully to predict the positions of electronic transitions in octahedral and tetrahedral complexes. The graphs are presented in parametric form with the ground term energy arbitrarily taken as zero. The parameters B and C may be obtained directly from atomic spectra. They are related to ion term splittings. Generally, B is found to be somewhat reduced in the complex from what it is in the free ion. The chemical implications of this reduction will be discussed later.

Let us assume that the Tanabe–Sugano diagram given in Figure 5-3 is appropriate for describing the electronic transitions of $Cr(OH_2)_6^{3+}$. The central ion, Cr^{III}, has a [core]$3d^3$ configuration and a ^4F ground state. Since B is ~ 1000 cm^{-1} for Cr^{3+} and $\Delta \sim 20$ kK for aqueous complexes of trivalent metal ions, a vertical line may be drawn

---

* Y. Tanabe and S. Sugano, *J. Phys. Soc. Japan* **9**, 753 (1954).

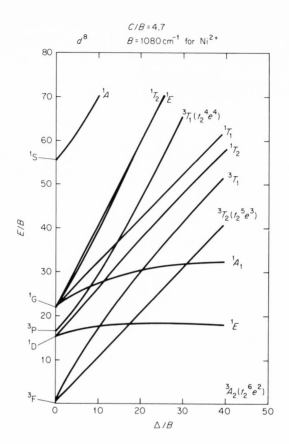

Figure 5-6   *Tanabe–Sugano energy level diagrams for the* d⁸ *ion in a cubic* (O_h) *field.*

parallel to the E/B ordinate from the number 20 on the abscissa. This line intersects the curves labeled $^4T_2(t_2^2e)$ at ∼20, $^4T_1$ near 27, and $^4T_1(t_2e^2)$ near 45. Since transitions to $^4T_2$, $^4T_1$, and $^4T_1(P)$ (this term comes from ^4P) require no spin change from that of the $^4A_2$ ground state, these will be the "spin-allowed" transitions. Such transitions are generally an order of magnitude more intense (more light is absorbed for a given concentration through a specific path length of sample) than the transitions requiring a spin change. The "spin-forbidden" transitions to doublet states such as ^2E or $^2T_1$ are much less readily observed. When found, however, their positions are fairly well predicted by the diagrams.

Experimentally, aqueous Cr^{III} shows three strong bands in the visible and low-energy ultraviolet (near u.v.) spectral region at 17.4, 24.6, and 37.8 kK. These bands must be associated with transitions from the $^4A_2$ ground state to the $^4T_2$, $^4T_1$, and $^4T_1(t_2e^2)$ states, respectively. To obtain a good fit of Figure 5-3 to the absorption spectrum, the crystal field parameter Δ becomes 17.4 kK. The inherent assumption that six water molecules surround the metal ion in solution in an octahedral* structure seems justified from these spectral results.

Exercise 5-1   At B equal to 960 cm^{-1} predict, by means of the Tanabe–Sugano diagram for the d^5 configurations, the energies of the sextet–quartet transitions for $Mn(OH_2)_6^{2+}$ assuming $\Delta = 8000$ cm^{-1}.

Exercise 5-2   Electronic transitions with molar absorbances of less than 100 are observed for $Ni(OH_2)_6^{2+}$ at 25,300 cm^{-1}, 13,500 cm^{-1}, and 8500 cm^{-1}. Assign these bands.

As indicated earlier in this chapter, the crystal field model places the t_{2g} orbitals in an octahedral complex at a lower energy than the e_g orbitals. In a tetrahedral complex, however, the d_{xz}, d_{yz}, and d_{xy} atomic orbitals (t_2) interact more strongly with ligands than do the $d_{x^2-y^2}$ and d_{z^2} atomic orbitals (e) so that the t_2 orbital is expected to be at a *somewhat higher energy* than the e orbital. Experimentally, the separation between t_2 and e is approximately one-half [based on electrostatic model $\Delta(T_d) = 4/9\Delta(O_h)$] as large in a tetrahedral molecule as it is in an octahedral complex with similar ligands.

The absence of the center of symmetry in T_d causes crystal field d–d transitions (such as $^4A_2 \to {}^4T_2$) in tetrahedral complexes to be more intense than d–d transitions in octahedral species. In a tetrahedral geometry p and d functions on the metal ion are no longer strictly separable. The orbitals d_{xz}, d_{yz}, d_{xy} and p_x, p_y, p_z all belong to the irreducible representation t_2. Thus electronic transitions labeled d $\to$ d display some d $\to$ p character and begin to take on some of the intensity that would be associated with the LaPorte allowed d $\to$ p atomic transitions (Chapter 2). However, unless the energies of the p and d orbitals are very close, we may assume the effect is small.

---

* While the point group symmetry of $Cr(OH_2)_6^{3+}$ is not strictly O_h, the dominant electrostatic effect on the metal ion produced by the water ligands is the same as if the six water molecules were replaced by negative charges. In general, the electrostatic field produced by those ligand atoms directly bonded to the metal ion controls the electronic structure with which we are concerned in the crystal field theory of coordination compounds.

CRYSTAL FIELD STABILIZATION ENERGY (CFSE)

The value of Δ or 10Dq can be a measure of the stabilization observed for a given geometrical structure over the hypothetical configuration in which the ligands are associated with the metal but with no specified geometry; that is, the metal–ligand coulombic interaction is the same in all directions. Since a completely filled set of hydrogenlike d orbitals has spherical symmetry, there can be no ligand field stabilization when the metal ion has a filled shell configuration. No stabilization can be expected for half-filled shells either (one electron in each orbital) without a pairing of the electron spins. Thus, neither Zn^{II} [core]$3d^{10}$ nor high-spin ("spin-free") Mn^{II} [core]$3d^5$ can exhibit a ligand field stabilization in their complexes.

For octahedral complexes the energy difference between e_g and t_{2g} orbitals, $[E(e_g) - E(t_{2g})]$, is Δ or 10Dq. Since for $(t_{2g})^6(e_g)^4$, the d^{10} configuration, there is no symmetry stabilization, we recognize that $6E(t_{2g}) + 4E(e_g) = 0$. As a result, the energy of the t_{2g} orbital becomes -4Dq, while that of the e_g orbital is $+6$Dq. Thus an octahedral complex with the metal ion configuration $(t_{1g})^6(e_g)^1$ has a crystal field stabilization* of -18Dq (Figure 5-7).

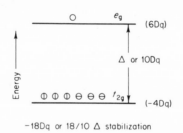

-18Dq or $18/10$ Δ stabilization

Figure 5-7 *Crystal field stabilization energy of a low-spin* d^7 *ion.*

* F. A. Cotton, *J. Chem. Ed.* **41**, 466 (1964) has described another way in which the concept of crystal field stabilization can be introduced. The total d-orbital population increases regularly from Ca^{2+} to Zn^{2+}. However, in ligand fields the different sets of d orbitals (the t_{2g} or e_g orbitals, for example) have separate populations which do not increase uniformly. The e_g population does go from 0 to 4. If this increase were uniform, a change of $\frac{4}{10}$ of an electron would occur from one ion to the next. Similarly, a change of $\frac{6}{10}$ of an electron would occur with uniform filling of the t_{2g} orbitals. With Mn^{2+}, five steps from Ca^{2+}, there would be two e_g electrons and three t_{2g} electrons, as found experimentally. The differences between the energy associated with "uniform" population and actual population is the CFSE. Thus for Cr^{3+}, the uniform populations are $(e_g)^{12/10}$

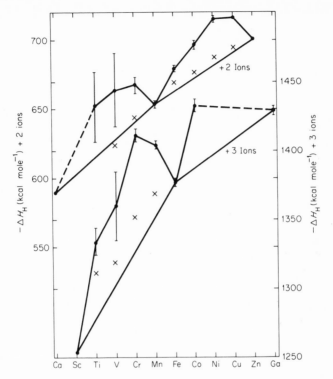

Figure 5-8 *The heat of hydration, ΔH (kcal), for the bivalent first-row transition series ions. Key: (●) experimental; (×) corrected. [After P. George and D. S. McClure, Prog. Inorg. Chem. 1, 418 (1959).]*

The CFSE is expected to influence the amount of energy released or absorbed in chemical reactions. An example is the equilibrium reaction:

$$Ni(NH_3)_6^{2+} + 6\ H_2O(l) \rightleftharpoons Ni(H_2O)_6^{2+} + 6\ NH_3(aq.) \tag{5-1}$$

According to the spectrochemical series, the CFSE of $Ni(NH_3)_6^{2+}$ is greater than that of $Ni(H_2O)_6^{2+}$. We might expect to see the influence of this energy on the degree to which the reaction proceeds to the right. However,

---

and $(t_{2g})^{18/10}$, while the actual populations are $(e_g)^0$ and $(t_{2g})^3$. The difference is the energy associated with removing $\frac{12}{10}$ electrons from e_g to t_{2g}. Since the separation of e_g and t_{2g} is 10Dq, the CFSE is $-(12/10) \times 10Dq = -12Dq$. The negative sign is given because a stabilization (decrease in energy) is implied; Dq is considered positive.

Table 5-5 *Crystal field stabilization energy* [a]

| Number of | Symmetry | |
d electrons	Octahedral	Tetrahedral
0	0	0
1	$-4Dq_o$	$-6Dq_t$
2	$-8Dq_o$	$-12Dq_t$
3	$-12Dq_o$	$-8Dq_t$
4	$-6Dq_o$	$-4Dq_t$
5	0	0
6	$-4Dq_o$	$-6Dq_t$
7	$-8Dq_o$	$-12Dq_t$
8	$-12Dq_o$	$-8Dq_t$
9	$-6Dq_o$	$-4Dq_t$
10	0	0

[a] Assuming maximum spin multiplicity.

since the CFSE is small compared with other energy releasing or absorbing processes implied in (5-1) (such as solvation of the released NH_3, the formation of six Ni—OH_2 bonds, the breaking of six Ni—NH_3 bonds, etc.), the CFSE associated with this reaction is not usually experimentally detectable.

$$M^{2+}(g) + 6\ H_2O(l) \leftrightharpoons M(H_2O)_6^{2+} \text{(solution)} \qquad (5\text{-}2)$$

The heat released upon dissolving a gaseous metal ion in water [Eq. (5-2)]—the negative of the heat of hydration, ΔH—is related to the CFSE of the hydrated complex. By plotting ΔH against the atomic number of the divalent first-row transition series ions (Figure 5-8) we might expect that the heat liberated would increase monotonously or smoothly as the ion size decreased from Ca to Zn. This is not the case; instead, a "double-humped" curve is obtained with maxima near V and Ni^{2+}. Taking into account the CFSE (Table 5-5) of each ion and determining Δ spectroscopically, a "correction" to the double-humped curve can be made which brings each point nearer to the dashed line.

SPECTROCHEMICAL SERIES

The spectroscopically determined value of Δ or $10Dq$ for a particular transition metal complex depends on the ion, its oxidation state, the ligands bonded to the metal, and variables such as pressure, temperature, etc. However, the dominant effects are due to the ligands, the particular metal, and its oxidation state. For a given metal ion, Δ has been observed to increase regularly within the ligand series $I^- < Br^- < Cl^- < H_2O < F^- < NH_3$. Such a list is called a "spectrochemical series." As the CFSE depends directly on Δ, it also increases for the above series.

Exercise 5-3 Calculate the CFSE in kcal mole^{-1} for $Ni(H_2O)_6^{2+}$, $\Delta = 8900$ cm^{-1}; $Co(H_2O)_6^{2+}$, $\Delta = 7560$ cm^{-1}; and $CrCl_6^{3-}$, $\Delta = 13,920$ cm^{-1}.

GROUND STATE SPIN CHANGES

As is apparent from the Tanabe–Sugano diagrams for cubic complexes (Figures 5-3 to 5-6), the ground state for some ions changes with increasing Δ. For the cubic d^1, d^2, d^8, and d^9 configurations, only one ground state is possible for all values of the crystalline field. For octahedral d^3 and tetrahedral d^7, this is also true. Other configurations, however, have more than one ground state. With octahedral d^4, for example, either $^5E_g(t_{2g}^3 e_g^1)$ or $^2T_1(t_2^4)$ is the ground state. The position of *spin-crossover* occurs here when Δ just balances the electron-repulsion energy associated with the placement of four electrons into t_2.

Exercise 5-4 Calculate the crystal field stabilization energy in units of Dq_0 for *low-spin* octahedral complexes of the d^4, d^5, d^6, and d^7 electron configurations.

LOWERING OF SYMMETRY

As we have already observed, a lowering of symmetry leads to a decrease in the degeneracy of orbital functions. What change might we expect for the electronic transitions in $Co(NH_3)_6^{3+}$ when two *trans* ammonia ligands are replaced by bromide ions to form *trans*-$Co(NH_3)_4Br_2^+$? This product has $\sim D_{4h}$ symmetry. An examination of the dimensions of the irreducible representations in this point group (Table 3-4) tells us immediately that the T terms must lose some of their degeneracy. The splitting to be observed will depend on the difference between the effective Δ for bromide compared with ammonia. With the

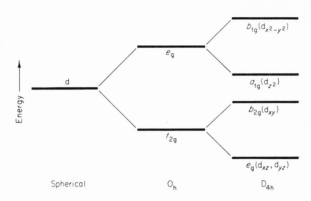

Figure 5-9 *Splitting of orbital levels in a tetragonal complex.*

bromide ions along the z axis in D_{4h}, the $d_{x^2-y^2}$ and d_{z^2} orbitals belong to the irreducible representations B_{1g} and A_{1g}, respectively, and are no longer degenerate. Also, d_{xz} and d_{yz} belong to E_g, but d_{xy} belongs to B_{2g}. The orbitals separate, as indicated in Figure 5-9, in proportion to the effective *tetragonality* of the complex. A completely tetragonal complex would be square planar, that is, no axial ligands are present. [If the axial ligands have a larger average Δ than the equatorial ones, the $b_{1g}(d_{x^2-y^2})$ orbital can be more stable than $a_{1g}(d_{z^2})$, while the $b_{2g}(d_{xy})$ orbital becomes more stable than $e_g(d_{xz}, d_{yz})$.]

To determine in a direct way what new electronic states result from a given symmetry state upon a lowering of symmetry, a straightforward application of group theory may be followed. In the above case, for example, we begin by recognizing that D_{4h} is a subgroup of O_h. Further, since we are concerned only with d orbitals, we may work simply with D_4 and add the subscript g later. Using only the operations of D_4, the irreducible representation for T_2 in O_h gives the reducible representation presented in Table 5-6. This can be reduced to $E + B_2$. Similarly, octahedral E becomes $A_1 + B_1$ in D_4. Thus, the $^5T_{2g}$ state in $Co(NH_3)_6^{3+}$

Table 5-6 *Octahedral irreducible representations in D_4*

D_4	E	$2C_4$	$C_2(C_4^2$ in $O_h)$	$2C_2'$	$2C_2''$
Γ_{T_2}	3	-1	-1	-1	1
Γ_E	2	0	2	0	0

splits up into the states ^5E_g and $^5B_{2g}$ on forming *trans*-Co$(NH_3)_4Br_2^+$. The spin multiplicity does not change, of course, unless the splitting is large enough to cause a spin-pairing. A convenient listing of the terms arising from O_h on reduction of symmetry to other important point groups is presented in the "correlation table" (Table 5-4).

LIGAND FIELD THEORY

It was stated before that the crystal field theory does not consider electron sharing between the metal ion and ligands. Clearly, electron sharing must happen in "real" compounds. The orbitals that we use to describe the bonding cannot be pure atomic orbitals. They are only partly metal atom in character as a result of metal–ligand bond formation. The consequences of bonding will become more apparent when we consider the molecular orbital model. We go part way toward this description in what follows.

Assuming that Δ is a spectroscopic parameter roughly related to the differences in bonding for a given metal ion to members of a series of ligands, and further, that the parameter B (the Racah parameter) which describes term separations in the free ion is an "adjustable" parameter in the complex, we have a *ligand field theory*. To use this theory, we follow the same techniques described above for crystal field theory. We recognize, however, that covalency in the bonding influences both Δ and B. In effect, we find a two-parameter fit to the Tanabe–Sugano diagrams. While the parameter Δ has been described sufficiently already, some discussion about the meaning of B is necessary.

NEPHELAUXETIC SERIES

C. K. Jørgensen has related the parameter β, defined by

$$\beta = \frac{B \text{ (complex)}}{B \text{ (free ion)}} \tag{5-3}$$

to a species-dependent series analogous to the spectrochemical series. This listing of elements or ligands according to a decreasing value of β is called a *nephelauxetic* (or "cloud-expanding") series. The parameter β appears to measure the *orbital diffusivity* of the metal–ligand bond and in a sense its covalent character. Values of β may range from $\sim$0.4 in a complex such as $Co(CN)_6^{3-}$ (which is generally considered very "covalent") to $\sim$0.95, as in the "ionic" MnF_6^{4-}. For some common ligands the nephelauxetic effect increases in the following order:

$$F^- < H_2O < NH_3 < NCS^- < Cl^- < CN^- < Br^- < I^-.$$

An alternative way to ascribe chemical significance to β is found in the "hard" and "soft" acid–base concept of Pearson.* The softness of a ligand corresponds roughly with its nephelauxetic effect. Hard ligands produce a low orbital diffusivity, hence large β values, while soft ligands give small β values. This concept also appears to be related to the ease with which a charge cloud or orbital of an atom is deformed by other charged species, that is, the polarizability of the atom. For example, the small, "hard" fluoride ion is not readily polarizable by a positive charge, so β is large; while the charge cloud of the large, "soft" iodide ion I^- is readily distorted by a cation, and β is small.

Table 5-7 *Spectrochemical parameters^a for Δ and β*

Ligands^b	f	h	Ions	g (cm^{-1})	k
$4NCS^-$	0.53	2.78	Co^{II}	7,560	0.09
$4Br^-$	0.42	2.74	Mn^{II}	8,000	0.07
$4Cl^-$	0.43	2.15	Ni^{II}	8,900	0.12
$4N_3^-$	0.52	3.55	Fe^{III}	14,400	0.24
$4OCN^-$	0.55	2.83			
$6H_2O$	1.00	1.0	Cr^{III}	17,400	0.21
$6NH_3$	1.25	1.4	Co^{III}	19,000	0.35
3en	1.28	1.5	Rh^{III}	27,000	0.30
3ox	0.98	1.4	Ir^{III}	32,000	0.3
$6Cl^-$	0.80	2.0			
$6CN^-$	1.7	2.0			
$6Br^-$	0.76	2.3			

^{a} $\Delta = fg$; $\beta = 1 - hk$.
^{b} en $= NH_2CH_2CH_2NH_2$; ox $= O_2CCO_2^{2-}$.

* R. G. Pearson, *J. Am. Chem. Soc.* **85**, 3533 (1963); *Science* **151**, 172 (1965).

Jørgensen has expressed the average values for Δ and β for transition metal complexes in terms of two ligand parameters, f and h, and two metal ion parameters, g and k:

$$\Delta = fg \; (\text{cm}^{-1}), \qquad \beta = 1 - hk \qquad (5\text{-}4)$$

In this way a large number of Δ and β values can be estimated (Table 5-7).

MOLECULAR ORBITAL THEORY

As stated by B. N. Figgis* "the chief feature we draw from molecular orbital theory [as applied to transition metal complexes] is the flexibility gained in the description of the t_{2g} and e_g (or t_2 and e) molecular orbitals." In the crystal field or ligand field theory the perturbing influence of the ligands on the metal orbitals is paramount, and the interaction of the ligand orbitals with each other as well as their specific interaction with the metal ion are not treated. By considering especially these latter interactions, we emphasize the fact that the d electrons in a transition metal complex are not only confined to the region of space near the metal ion, but actually become associated with the ligands. If we construct a molecular orbital for an octahedral complex so that each ligand atom has $\frac{1}{12}$ of a unit of charge density and the metal ion has $\frac{6}{12}$ of a unit when we put an electron in this orbital, we would say that the electron is shared equally between the ligands and the metal, that is, we have covalent bonding. While this perfect sharing of electrons is no more realistic than the complete lack of electron sharing implied by the crystal field theory, the molecular orbital approach does represent a fairly successful attempt to describe the bonding.

Exercise 5-5 With the help of the Tanabe–Sugano diagrams and the spectrochemical parameters of Table 5-7, estimate the expected positions for the spin-allowed d–d transitions in $Co(N_3)_4^{2-}$, $Co(OH_2)_6^{3+}$, $FeCl_4^-$, $Rh(NH_3)_6^{3+}$, and $Cr(ox)_3^{3-}$.

* B. N. Figgis, "Introduction to Ligand Fields," p. 201. Wiley (Interscience), New York, 1966.

A molecular orbital for our purposes is a linear combination of atomic orbitals (LCAO–MO). We form a $\sigma(1s)$ bonding orbital for H_2 by combining the 1s orbitals, ψ_{1s}, on each of the hydrogen atoms as in

$$\psi_{MO} = N[\psi_{1s(A)} + \psi_{1s(B)}] \tag{5-5}$$

In the molecular orbital, ψ_{MO}, N is a normalization constant which causes the probability of finding an electron in ψ_{MO} to be unity (or perfect) when all possible regions of space associated with ψ_{MO} are explored. This relationship is expressed mathematically by

$$\int_0^\infty [\psi_{MO}]^2 \, d\tau = 1 \tag{5-6}$$

The integration over $d\tau$ from 0 to $+\infty$ signifies the consideration of all space.*

The value of the normalization constant for the ψ_{MO} of Eq. (5-5) is obtained by expansion and integration:

$$N = \left[\int_0^\infty [\psi_{1s(A)} + \psi_{1s(B)}]^2 \, d\tau \right]^{-1/2} \tag{5-7}$$

This equation is simplified to

$$N = \left[2 \int_0^\infty \psi_{1s}^2 \, d\tau + 2S \right]^{-1/2} = 2^{-1/2}(1 + S)^{-1/2} \tag{5-8}$$

if the atomic orbitals themselves are normalized, where the overlap integral, S, is given by

$$S = \int_0^\infty \psi_{1s(A)}\psi_{1s(B)} \, d\tau \tag{5-9}$$

---

* Since wave functions may be complex, Eq. (5-6) should read

$$\int_0^\infty \psi_{MO}^*\psi_{MO} \, d\tau = 1$$

where ψ_{MO}^* is the complex conjugate of ψ_{MO}. As we generally are concerned only with real wave functions, the asterisk usually will not be specified.

Since S usually is small compared with one, $N = 1/\sqrt{2}$ and the LCAO–MO wave function for ψ_{MO} is given by

$$\psi_{MO} = \frac{1}{\sqrt{2}} \left[\psi_{1s(A)} + \psi_{1s(B)} \right] \tag{5-10}$$

ORBITAL OVERLAP

The overlap integral [Eq. (5-9)] is itself an important quantity since it tells us how much sharing of space actually is involved in our molecular orbital. Clearly, the greater the value of S, the more $\psi_{1s(A)}$ interacts with $\psi_{1s(B)}$. Of course, when $\psi_{1s(A)}$ becomes identical to $\psi_{1s(B)}$, S becomes its maximum, one. Symmetry tells us something about S, since S must be identically zero when $\psi_{1s(A)}$ and $\psi_{1s(B)}$ belong to different (orthogonal) irreducible representations; $\Gamma(\psi_{1s(A)}) \Gamma(\psi_{1s(B)})$ is zero. *Only orbitals of the same symmetry can overlap.* In an octahedral complex we will find that the e_g orbitals in the metal have the proper symmetry to interact with a LCAO–MO formed from ligand atom orbitals directed along the metal—ligand bonds (sigma bonds). The t_{2g} set of metal orbitals interact with an LCAO–MO formed from ligand orbitals which are perpendicular (pi bonds) to the metal—ligand bond.

CONSTRUCTION OF LCAO–MO

Character tables can be used to construct LCAO–MOs which conform to the symmetry of the molecule. For H_2 as an example, the molecular orbital formed from a linear combination of 1s atomic orbitals on each atom must be consistent with the $D_{\infty h}$ symmetry of the molecule. If we represent the 1s orbitals as in Figure 5-10, the reducible representation in $D_{\infty h}$ is given by Table 5-8. We see this is true since $1s(A)$ and $1s(B)$ each remain unchanged by E, C_{∞}^{ϕ}, and σ_v, while they are carried into each other by i, S_{∞}^{ϕ}, and C_2 (Table 5-9). By inspection (see Table 5-10) this representation reduces to $\sum_g^+$ (sigma g plus) plus $\sum_u^+$. Thus the two LCAO–MOs formed from $1s(A)$ and $1s(B)$ belong to the symmetry representations* $\sum_g^+$ and $\sum_u^+$.

---

* With linear molecules, symmetry notations historically are given by Σ, π, and Δ; Σ notation indicates one-dimensional representation, while π and Δ are two-dimensional. The $+$ sign indicates that the representation is symmetric with respect to reflections in a mirror containing the molecule.

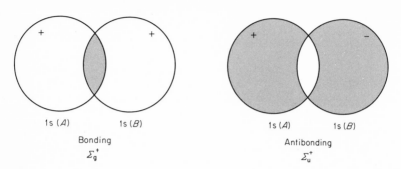

Figure 5-10 *Overlap of hydrogen 1s orbitals in bonding and antibonding LCAO–MOs.*

In general, the construction of LCAO–MOs begins by considering how each basis orbital transforms under the operations of the point group in question. For hydrogen, we see that $1s(A)$ (labeled A) and $1s(B)$ (labeled B) transform in $D_{\infty h}$, as given in Table 5-9 (it is sufficient to consider only one σ_v and one C_2). From a reduction of the reducible representation, Γ_r, one of the LCAO–MO symmetry orbitals must be invariant (Σ_g^+) to each of the operations of $D_{\infty h}$. This MO is generated by using the numbers in Σ_g^+ under each operation as a coefficient for the term appearing in the transformation table under the same operation. For example, combining the characters in Σ_g^+ with A we have $A + A + A + A + B + B + B + B$ or $4(A + B)$. Similarly, for B we obtain the same result, $4(A + B)$. Ignoring the multiplicative constant, we see that the function $(A + B)$ is Σ_g^+ in the point group $D_{\infty h}$. For Σ_u^+, the functions $4(A - B)$ and $-4(A - B)$ are obtained. Since we are concerned at present only with the symmetry properties of these functions, the coefficient ± 4 can be ignored. The LCAO–MO generally is normalized to unity anyway. Doing this by neglecting the overlap integral, we arrive at the

Table 5-8 *The reducible representation for hydrogen atom 1s orbitals in the $D_{\infty h}$ symmetry of H_2*

$D_{\infty h}$	E	$2C_\infty^\phi \cdots \infty \sigma_v$	i	$2S_\infty^\phi \cdots \infty C_2$
Γ_r	2	$2 \cdots 2$	0	$0 \cdots 0$

Table 5-9 *Transformations of 1s(A) and 1s(B) for hydrogen in* $D_{\infty h}$

$D_{\infty h}$	E	C_∞^ϕ a	C_∞^ϕ b	$\infty \sigma_v$	i	S_∞^ϕ a	S_∞^ϕ b	∞C_2
A	A	A	A	A	B	B	B	B
B	B	B	B	B	A	A	A	A

a Clockwise rotation.
b Counterclockwise rotation.

anticipated result:

$$\psi(\textstyle\sum_g^+) = \frac{1}{\sqrt{2}}(A + B), \quad \psi(\textstyle\sum_u^+) = \frac{1}{\sqrt{2}}(A - B) \quad (5\text{-}11)$$

Until now we have not suggested which sigma molecular orbital, $\sum_g^+$ or $\sum_u^+$, will be most stable and hence contain the two electrons of H_2 in its ground state. The relative energies of the two orbitals can be determined by solving the Schrödinger equation

$$\mathcal{H}\psi(\textstyle\sum_g^+) = \epsilon_1\psi(\textstyle\sum_g^+), \quad \mathcal{H}\psi(\textstyle\sum_u^+) = \epsilon_2\psi(\textstyle\sum_u^+) \quad (5\text{-}12)$$

for each of the MOs. To do this we multiply each side of the equation by

Table 5-10 *Character table for* $D_{\infty h}$

$D_{\infty h}$	E	$2C_\infty^\phi$	$\cdots$	$\infty \sigma$	i	$2S_\infty^\phi$	$\cdots$	∞C_2		
Σ_g^+	1	1	$\cdots$	1	1	1	$\cdots$	1		$x^2 + y^2, z^2$
Σ_g^-	1	1	$\cdots$	-1	1	1	$\cdots$	-1	R_z	
H_g	2	$2\cos\phi$	$\cdots$	0	2	$-2\cos\phi$		0	(R_x, R_y)	(xz, yz)
Δ_g	2	$2\cos 2\phi$	$\cdots$	0	2	$2\cos 2\phi$	$\cdots$	0		$(x^2 - y^2, xy)$
$\vdots$	$\vdots$	$\vdots$	$\cdots$	$\vdots$	$\vdots$	$\vdots$	$\cdots$	$\vdots$		
Σ_u^+	1	1	$\cdots$	1	-1	-1	$\cdots$	-1	z	
Σ_u^-	1	1	$\cdots$	-1	-1	-1	$\cdots$	1		
H_u	2	$2\cos\phi$	$\cdots$	0	-2	$2\cos\phi$	$\cdots$	0	(x, y)	
Δ_u	2	$2\cos 2\phi$	$\cdots$	0	-2	$-2\cos 2\phi$	$\cdots$	0		
$\vdots$	$\vdots$	$\vdots$	$\cdots$	$\vdots$	$\vdots$	$\vdots$	$\cdots$	$\vdots$		

the wave function and integrate over all space.* For $\sum_g^+$ we have

$$\int_0^\infty \psi(\textstyle\sum_g^+)\,\mathcal{3C}\psi(\textstyle\sum_g^+)\,d\tau = \epsilon_1 \int_0^\infty [\psi(\textstyle\sum_g^+)]^2\,d\tau \qquad (5\text{-}13)$$

If we neglect the overlap integral S, the right-hand side of (5-13) becomes simply ϵ_1, the energy associated with $\sum_g^+$. To evaluate the left-hand side of (5-13) it is necessary to know $\mathcal{3C}$, the Hamiltonian for the system. In practice this is difficult since a good *ab initio* Hamiltonian leads to equations which are not presently solvable, at least for complex molecules. However, we can expand the left-hand side of (5-13) and set the resulting integrals equal to parameters, as in

$$\int (A + B)\mathcal{3C}(A + B)\,d\tau$$

$$= \int A\mathcal{3C}A\,d\tau + \int B\mathcal{3C}B\,d\tau + \int A\mathcal{3C}B\,d\tau + \int B\mathcal{3C}A\,d\tau$$

$$= 2\alpha + 2\beta \qquad (5\text{-}14)$$

The integrals $\int A\mathcal{3C}A\,d\tau$ and $\int B\mathcal{3C}B\,d\tau$ are identical since the atoms are identical. This type of integral is called a *coulomb integral*, α, and represents the energy of the individual atomic basis function. The $\int A\mathcal{3C}B\,d\tau$ and $\int B\mathcal{3C}A\,d\tau$ integrals also have the same value and are called *resonance integrals*, β. They describe the bonding interaction between the atomic orbitals. The energies for $\epsilon_1(\sum_g^+)$ and $\epsilon_2(\sum_u^+)$ are

$$\epsilon_1\left(\textstyle\sum_g^+\right) = \alpha + \beta, \qquad \epsilon_2\left(\textstyle\sum_u^+\right) = \alpha - \beta \qquad (5\text{-}15)$$

The integrals α and β are negative by convention, making the $\sum_g^+$ orbital more stable than the $\sum_u^+$ orbital. Since there is no node in the $\sum_g^+$ function, it describes a bonding interaction. The energy separation between the two LCAO–MOs is 2β.

Exercise 5-6   The ion $Re_2Cl_8^{2-}$ has D_{4h} symmetry (see Figure 1-2). By using the $d_{x^2-y^2}$, s, p_x, and p_y orbitals on each Re^{3+} ion to form sigma bonds to the chlorides, construct

---

* Again we must remember to use complex conjugates when imaginary functions are involved.

appropriate LCAO–MO orbitals for the remaining metal d orbitals in the complex. Determine the relative energies of the MOs assuming

$$\int d'_{z^2}\Im c d_{z^2}\, d\tau < \int d'_{xy}\Im c d_{xy}\, d\tau$$

$$< \int d'_{xz}\Im c d_{xz}\, d\tau < \int p'_{z}\Im c p_{z}\, d\tau.$$

Assume the coulomb integrals are equal for all orbitals.

GROUP ORBITALS IN SIGMA BONDING

Consider now the LCAO–MO description of the bonding in a square planar molecule (ignoring protons) such as $Pt(NH_3)_4^{2+}$. We will assume that the only interaction between Pt^{II} and the NH_3 ligands occurs between metal orbitals on Pt^{II}, which are hydrogenlike, and group orbitals on the ligands which are constructed from appropriate

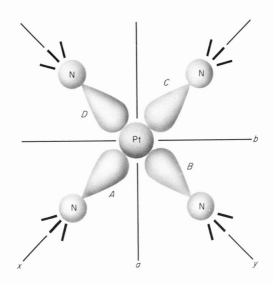

Figure 5-11 *Sigma bonding ligand orbitals in* $Pt(NH_3)_4^{2+}$ *used to form ligand group orbitals.*

Table 5-11   *Reducible representation for ligand sigma functions in* $Pt(NH_3)_4^{2+}$

D_{4h}	E	$2C_4$	C_2	$2C_2'$	$2C_2''$	i	$2S_4$	σ_h	$2\sigma_v$	$2\sigma_d$
Γ_r	4	0	0	2	0	0	0	4	2	0

linear combinations of ligand sigma orbitals.*The particular ligand orbital we consider on each nitrogen could be an sp³ hybrid function or simply an atomic s function, but this point is unimportant at present. The only requirement is that a ligand basis orbital is directed along the metal–ligand bond. The ligand orbitals (Figure 5-11) are labeled A, B, C, and D. The reducible representation to which they belong in the D_{4h} symmetry group is given in Table 5-11. On reduction of this representation we observe that these in-plane ligand orbitals span the irreducible representations A_{1g}, B_{1g}, and E_u. The metal orbitals with these same symmetries (Table 3-4) are the $s(A_{1g})$, $d_{z^2}(A_{1g})$, $d_{x^2-y^2}(B_{1g})$, and p_x, $p_y(E_u)$ functions. Since orbitals of the same symmetry interact providing their energies are similar, molecular orbitals can be formed by combining the ligand group orbitals with metal orbitals of the same symmetry.

To determine the form of the ligand group orbitals, an orbital transformation table is constructed. Since g's and u's can be added after the rotational symmetry of the group is determined, consider only how the ligand orbitals transform under the operations of D_4 (Table 5-12).

Table 5-12   *Transformation table for sigma orbitals in* $Pt(NH_3)_4^{2+}$

D_4	E	$C_4{}^a$	C_4	C_2	$C_2'(x)$	$C_2'(y)$	$C_2''(xy)$	$C_2''(xy)$
A	A	B	D	C	A	C	D	B
B	B	C	A	D	D	B	C	A
C	C	D	B	A	C	A	B	D
D	D	A	C	B	B	D	A	C

^{a} Counterclockwise rotation.

* Group orbitals are simply linear combinations of ligand atomic orbitals. The molecular obritals (LCAO–MOs) are created to conform to the symmetry of the complex as a whole. While calculations can be performed without first generating separate ligand group orbitals, the group orbital procedure allows for the easy identification of molecular orbitals in the complex which are largely either ligand- or metal-ion based.

To construct the ligand group orbitals, ψ_i, for $Pt(NH_3)_4^{2+}$ we follow the same procedure used for H_2. These orbitals must conform to the irreducible representations A_{1g}, B_{1g}, and E_u in D_{4h}. To make them conform, we use as coefficients the characters for the appropriate representations as found in the character table for D_{4h}. Since A_{1g} and B_{1g} are symmetric to inversion and E_u is asymmetric, we need use only operations and characters of D_4. Thus, $\psi_1(A_{1g})$ is formed from the characters of A_1, with the same result being obtained by using transformation properties (Table 5-12) of either A, B, C, or D. Normalization with neglect of overlap gives

$$\psi_1(A_{1g}) = \frac{1}{2}(A + B + C + D)$$

$$\psi_2(B_{1g}) = \frac{1}{2}(A - B + C - D)$$

(5-16)

$$\psi_3(E_u) = \frac{1}{\sqrt{2}}(A - C)$$

$$\psi_4(E_u) = \frac{1}{\sqrt{2}}(B - D)$$

Similarly, the other symmetry group orbitals are obtained. (It should be noted that $(A - C)$ and $(C - A)$ only differ by a multiplicative constant.) With the help of Figure 5-11, it is readily seen that ψ_3 and ψ_4 are *ungerade* to inversion, while ψ_1 and ψ_2 are *gerade*.

Complete LCAO–MOs are formed by combining appropriate metal orbitals with the ligand group orbitals of the same symmetry (Figure 5-12). Unless the electrons are shared equally by the metal and the ligands ($a_i = b_i$), the *mixing coefficients* a_i and b_i will be different. Their magnitude depends on the particular ligand and metal orbitals involved. If we assume that the overlap integrals are small compared with unity, $b_i = \pm(1 - a_i^2)^{1/2}$. In transition metal coordination compounds b_i generally is larger than a_i for bonding functions and the MO is primarily ligand in character. For antibonding functions (Figure 5-13), the resulting MOs are largely metal type orbitals. These antibonding functions are like the crystal field or ligand field perturbed metal orbitals we discussed previously.

Exercise 5-7   Construct group orbitals for the sigma bonds in tetra-hedral $FeCl_4^-$. What metal orbitals interact with these ligand orbitals?

$$\psi_1(A_{1g}) = a_1 S + \frac{b_1}{2}(A + B + C + D)$$

$$\psi_2(B_{1g}) = a_2 d_{x^2-y^2} + \frac{b_2}{2}(A - B + C - D)$$

$$\psi_3(E_u) = a_3 p_x + \frac{b_3}{\sqrt{2}}(A - C)$$

$$\psi_4(E_u) = a_3 p_y + \frac{b_3}{\sqrt{2}}(B - D)$$

Figure 5-12   LCAO–MO *bonding sigma orbitals for* $Pt(NH_3)_4^{2+}$ *(neglecting overlap).* *(Since* d_{z^2} *also belongs to* A_{1g} *in* D_{4h}, *it normally will be part of the* A_{1g} *wave function. However, we will assume here that its contribution to* $\psi_1(A_{1g})$ *is negligibly small.)*

$$\psi_1^*(A_{1g}) = b_1 S - \frac{a_1}{2}(A + B + C + D)$$

$$\psi_2^*(B_{1g}) = b_2 d_{x^2-y^2} - \frac{a_2}{2}(A - B + C - D)$$

$$\psi_3^*(E_u) = b_3 p_x - \frac{a_3}{\sqrt{2}}(A - C)$$

$$\psi_4^*(E_u) = b_3 p_y - \frac{a_3}{\sqrt{2}}(B - D)$$

Figure 5-13   LCAO–MO *antibonding sigma orbitals for* $Pt(NH_3)_4^{2+}$ *(neglecting overlap).*

LCAO–MO ENERGY LEVEL DIAGRAMS

A qualitative LCAO–MO energy level diagram taking into account only sigma bonding interactions is presented in Figure 5-14.

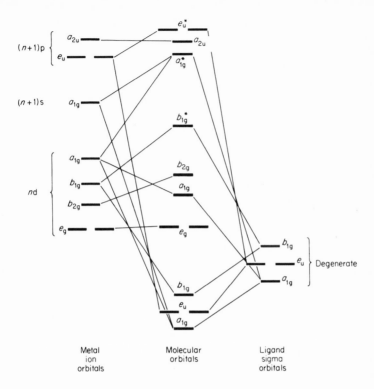

Metal
ion
orbitals

Molecular
orbitals

Ligand
sigma
orbitals

Figure 5-14   LCAO–MO *energy level diagram for a tetragonal complex* (D_{4h}) *with no pi bonding.*

For a d^8 metal ion such as Pt^{II} or Ni^{II}, the lowest-energy electronic transitions are from the nonbonding $b_{2g}(d_{xy})$ orbital to the antibonding b_{1g}^* orbital which, as indicated previously, is largely metal ion in character. A similar LCAO–MO diagram is presented in Figure 5-15 for an octahedral complex. The relationship between the ligand field parameter Δ and the separation between the antibonding t_{2g}^* orbitals and the antibonding e_g^* orbitals is apparent—the stronger the bond, the larger the value† of Δ.

† Caution should be exercised before assuming that Δ depends only on sigma bonding. In the LCAO–MO model, the metal t_{2g} orbitals are largely pi antibonding in character. Thus a strong pi bonding ligand may increase the energy of t_{2g} as well as e_g^*, giving a smaller Δ than expected based on sigma bonding alone. Similarly, empty pi orbitals on ligands may accept electrons from t_{2g} and lower its energy, thus increasing Δ substantially.

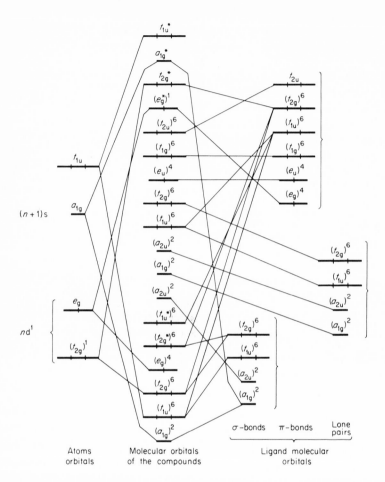

Figure 5-15 *Molecular orbitals for an octahedral complex with a d^1 configuration, including both sigma and pi bonding. [After A. D. Liehr, J. Chem. Ed. **39**, 135 (1962).]*

Exercise 5-8   Predict the number of unpaired electrons expected in a tetragonal complex of Ni^{2+}: (1) in the crystal field formalism; (2) with molecular orbital theory. What assumptions regarding the bonding must be made to make the result identical for each theory?

SUPPLEMENTARY READING

C. J. Ballhausen and H. B. Gray, "Molecular Orbital Theory." Benjamin, New York, 1965.

B. N. Figgis, "Introduction to Ligand Fields." Wiley (Interscience), New York, 1966.

R. Krishnamurthy and W. B. Schaap, *J. Chem. Ed.* **46,** 799 (1969).

For an advanced mathematical treatment of bonding in transition metal compounds, the reader is referred to J. S. Griffith, "The Theory of Transition Metal Ions." Cambridge Univ. Press, Cambridge, 1961.

6

The d-block transition elements and their chemistries

The descriptive chemistry of the transition elements is interestingly varied. Unlike the main group elements, transition elements in the same formal periodic group or triad often display considerably different chemistries. Including all transition elements, there are approximately fifty different chemistries to describe—something we will not attempt here. Among the d-block elements themselves, some important chemical similarities are found. Three fairly obvious ones are as follows:

(a) Except for Hg, the elements are all relatively hard, strong metals. They conduct electricity and heat, and easily form alloys with each other and other metals.

(b) In general, these elements dissolve in mineral acids with the production of H_2; the "noble" elements such as Pt, Au, Rh, Ir, Ru, Os, and Pd are exceptions to this rule.

(c) Nearly all transition elements form ions in a wide variety of formal oxidation states, most of which are colored. A large number of the ions formed are also paramagnetic.

The aqueous chemistry of the first transition series is considerably better understood than the aqueous chemistry of the second and third transition series. A major contributing factor to this state of affairs has been the applicability of ligand field theory to a description of the electronic properties of first transition series compounds. With the second and third transition series, the electronic structures are considerably more difficult to unravel due to larger metal ion spin–orbit coupling and increased metal–ligand atom covalency. Furthermore, dimerization and oligomerization (trimers, tetramers, etc.) often frustrate the successful description of the second and third transition species actually present in solution.

In this chapter the reader is introduced to the descriptive chemistry of a select few d-block transition elements and their compounds. Hopefully, sufficient interest in these elements will be generated to cause the student to explore in more depth some of the exciting science presently being uncovered with other transition metal compounds. For continued exploration the reader is referred to other texts and review articles.

FIRST TRANSITION SERIES

The elements forming ions with the electron configuration $[Ar](3d)^n$ belong to the first transition series. The elements themselves (Table 2-1) display a $3d^n4s^2$ structure, except for chromium and copper. Ionization of the s electrons generally leads to the formation of prominent oxidation states of II or higher (see Chapter 2). The relatively stable $3d^{10}$ configuration present in Zn^{II} also is present in Cu^I. A list of common oxidation states* and ionization potentials for the first transition series is given in Table 6-1.

Table 6-1 Common[a] oxidation states and ionization potentials of the first transition series elements

Element	Sc	Ti	V	Cr	Mn	Fe	Co	Ni	Cu	Zn
Oxidation states	III	III	II	II	II	II	II	II	I	II
		IV	III	III	III	III	III		II	
			IV	VI	VI					
			V							
Ionization	1st 6.56	6.83	6.74	6.76	7.43	7.90	7.86	7.63	7.72	9.39
potentials[b]	2nd 12.89	13.57	14.65	16.49	15.64	16.18	17.05	18.15	20.29	17.96

[a] Commonly found in aqueous solution.
[b] Electron volts.

* Oxidation states are so sensitive to environmental conditions that the word "common" needs definition. We use it here only to mean states having more than a "fleeting" existence in water.

SCANDIUM

While not a rare element ($\sim 10^{-7}$ percent of igneous rock), Sc is so widely distributed in nature that its chemistry has not been thoroughly studied due to a lack of supply (a price of \$60 per gram recently has been quoted for the metal). Except for gas-phase molecules such as ScO ($C_{\infty v}$), the Sc^{III} oxidation state is generally obtained by loss of the three outer electrons. The chemical behavior is much like that of aluminum except that the sesquioxide, Sc_2O_3, is more basic than Al_2O_3. It does dissolve readily in excess concentrated hydroxide solution (amphoteric) to form species which presumably are hydrated $Sc(OH)_6^{3-}$ or $ScO(OH)_4^{3-}$. The monomeric nature of the trihalide such as $ScCl_3$ (D_{3h}) distinguishes it from the dimeric aluminum halide, $[AlCl_3]_2$ (D_{2h}). Some reactions of scandium and its compounds are

$$Sc(s) + O_2(g) \rightarrow Sc_2O_3(s)$$

$$Sc(s) + HCl \text{ (dil)} \rightarrow ScCl_2^+, ScCl^{2+} + H_2 \uparrow$$

$$Sc^{3+} + OH^- \text{ (dil)} \rightarrow Sc_2O_3 \cdot nH_2O \downarrow$$

$$Sc_2O_3 \cdot nH_2O(s) + OH^- \text{ (conc)} \rightarrow Sc(OH)_6^{3-}$$

$$ScCl_3(aq) + HAcAc + base \rightarrow Sc(AcAc)_3 \downarrow$$

TITANIUM

All valence electrons are readily lost from Ti atoms to produce the Ti^{IV} oxidation state. The Ti^{III} and Ti^{II} states also are known, but their compounds generally are less stable. In the Ti^{IV} state, properties are similar to the Group IV ions, particularly Sn^{IV}, which has an ionic radius only a few hundredths of an angstrom larger than that of Ti^{IV} (0.68 Å).

The relatively abundant metal (~ 0.4 percent igneous rock as oxides such as TiO_2, *rutile*) has become important as a space age material due to its hardness, high melting point (1812°C), low density (4.5 gm cm^{-3}), and relative unreactivity. This latter property results from the formation of a strong surface coat of oxide or nitride which protects the metal from further oxidation. Hot hydrochloric acid or fluoride-containing mineral acids dissolve the metal to produce Ti^{III} species.

The tetrahalides of Ti^{IV} are fairly volatile, except for TiF_4, which may be ionic. They are all Lewis acids and readily react with excess halide or bases such as amines to form species of higher coordination numbers. The cationic $Ti(AcAc)_3^+$ is stabilized by anions such as $FeCl_4^-$ and presumably contains an octahedral coordination of O atoms about the metal with a D_3 symmetry (ignoring protons).

The presence of one 3d electron in Ti^{III} makes compounds of this oxidation state very interesting to study. Theory suggests that spin–orbit coupling should lead to magnetic moments (see Chapter 2, $L \neq 0$) at room temperature of ~ 1.86 B.M., but most Ti^{III} compounds display the "spin-only" moment of 1.73 B.M. This reduction in moment has been attributed to distortions of the complexes from cubic symmetry.

In water a violet Ti^{III} solution possesses an electronic absorption band at 20,300 cm⁻¹ with a shoulder at 17,400 cm⁻¹ (Figure 6-1). The absorption is attributed to the $^2T_{2g} \rightarrow {}^2E_g$ transition in an octahedral ligand field. Since the excited state ^2E_g arises from placement of an electron in one of two degenerate antibonding d orbitals, $d_{x^2-y^2}$ or d_{z^2}, a Jahn–Teller effect is produced which requires a splitting of the ^2E_g state. (Jahn–Teller effects are commonly found for $3d^4$ and $3d^9$ ground state configurations as in Cr^{II} and Cu^{II} complexes. The phenomenon will be discussed in some detail later.) The electronic spectrum is consistent with the presence of a $Ti(OH_2)_6^{3+}$ species in solution with essentially O_h

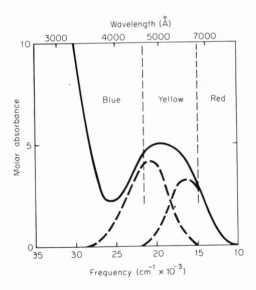

Figure 6-1 *The visible absorption spectrum of* $[Ti(H_2O)_6]^{3+}$.

symmetry. A similar species is present in the alum, $CsTi(SO_4)_2 \cdot 12H_2O$. Some reactions of titanium and its compounds are

$$TiO_2(s) + C(s) + Cl_2(g) \rightarrow TiCl_4(b.p.\ 137°C)$$

$$TiCl_4(l,\ m.p.\ -23°C) + 2H_2O \rightarrow TiO_2(s) + 4\ HCl$$

$$TiO_2 \cdot nH_2O + OH^-\ (conc) \rightarrow soluble\ oxides$$

$$Ti(ClO_4)_4 \cdot nH_2O + HF_2^- \rightarrow TiF_6^{2-}$$

CHROMIUM

The chemistry of Cr^{III} complexes generally is that of an "octahedral" species in which ligands are relatively slowly exchanged (several hours to days are required for exchange of approximately one-half of the ligands). H. Taube has described such species as "nonlabile" or inert. Chemical lability leads to rapid ligand scrambling. In nonlabile complexes (Table 6-2) displacements will occur but the rate-determining

Table 6-2 *Six-coordinate labile and inert complexes*

Configuration	Labile complexes
$t_{2g}^0 e_g^0$	Sc^{III}, Y^{III}, Ti^{IV}, Zr^{IV}, Hf^{IV}, Nb^V, Ta^V, Mo^{VI}, W^{VI}
$t_{2g}^1 e_g^0$	Ti^{III}, V^{IV}, Mo^V, W^V, Re^{VI}
$t_{2g}^2 e_g^0$	Ti^{II}, V^{III}, Nb^{III}, Ta^{III}, Mo^{IV}, W^{IV}, Re^V, Ru^{VI}
$t_{2g}^3 e_g^1$	Cr^{II}, Mn^{III}
$t_{2g}^3 e_g^2$	Mn^{II}, Fe^{III}
$t_{2g}^4 e_g^2$	Fe^{II}
$t_{2g}^5 e_g^2$	Co^{II}
$t_{2g}^6 e_g^2$	Ni^{II}
$t_{2g}^6 e_g^3$	Cu^{II}
$t_{2g}^6 e_g^4$	Zn^{II}, Cu^I, Ag^I, Cd^{II}, Hg^{II}

Configuration	Inert complexes
$t_{2g}^3 e_g^0$	V^{II}, Cr^{III}, Mo^{III}, W^{III}, Mn^{IV}, Re^{IV}
$t_{2g}^4 e_g^0$	$Cr(CN)_6^{4-}$, $Mn(CN)_6^{3-}$, Re^{IV}, Ru^{IV}, Os^V
$t_{2g}^5 e_g^0$	$Mn(CN)_6^{4-}$, Re^{II}, Ru^{III}, Os^{III}, Ti^{IV}
$t_{2g}^6 e_g^0$	$Mn(CN)_6^{5-}$, $Fe(CN)_6^{4-}$, Ru^{II}, Os^{II}, Co^{III}, Rh^{III}, Ir^{III}, Pd^{IV}, Pt^{IV}

processes generally involve considerable metal–ligand bond breaking energy. As a result of their kinetic stability, isomers such as violet $[Cr(H_2O)_6]Br_3$ and green $[Cr(H_2O)_4Br_2]Br \cdot 2H_2O$ are known. Small amounts of Cr^{II} can catalyze ligand exchange of Cr^{III} complexes via electron transfer reactions such as

$$Cr^{II^{2+}}(aq) + (NH_3)_5Cr^{III}Cl^{2+} \rightarrow [(NH_3)_5CrClCr(OH_2)_5^{4+}]$$

$$\rightarrow 5\ NH_3 + Cr^{II^{2+}}(aq) + ClCr^{III}(OH_2)_5^{2+} \qquad (6\text{-}1)$$

The atom transfer which results requires the intermediacy of bridged species such as suggested in (6-1). The aqueous Cr^{II} is kinetically reactive (labile) and rapidly exchanges ligands with the solvent. The particular Cr^{II} complexes present are controlled only by equilibrium considerations.

MANGANESE

At manganese the 3d shell becomes so stable that the 4s electrons are removed considerably more easily than the 3d electrons. The stability of the 3d shell causes the most important oxidation state in aqueous solutions to be Mn^{II} for each of the elements from manganese through zinc. Higher oxidation states become increasingly difficult to achieve.

Mineral acids such as $HClO_4$ oxidize manganese metal [forming $H_2(g)$] to produce solutions which are nearly colorless. In high concentration these solutions are generally pink, similar in color to salts containing $Mn(H_2O)_6^{2+}$. The ground state for the octahedral high-spin species is $^6A_{1g}$. Visible region electronic transitions are spin-forbidden and the molar absorbance is low. The values observed are approximately 1–2 orders of magnitude less than for comparable spin-allowed d–d bands. The spectrum of $Mn(H_2O)_6^{2+}$ (Figure 6-2) corresponds favorably with the predictions of ligand field theory for an octahedral complex.

Examination of the Tanabe–Sugano diagram for a d^5 ion (Figure 5-4) shows that a $^2T_{2g}$ ground state can be obtained for octahedral Mn^{II} complexes if Δ/B is large. Similarly, d^4 ions such as Cr^{II} or Mn^{III} produce either high-spin or low-spin cubic complexes, depending on the magnitude of Δ/B. For Mn^{II} complexes a simple calculation based on the information of Table 5-7 suggests that Jørgensen's f parameter must be near 3 for reasonable values of the ligand parameter h. This would lead to the conclusion that none of the ligands listed in the table produce

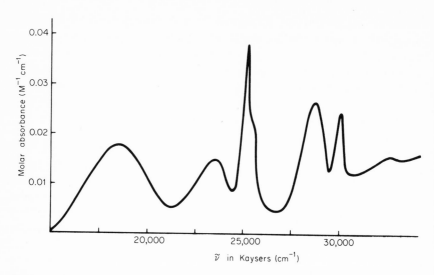

Figure 6-2   *Visible absorption spectrum of aqueous* Mn(II).

a low-spin complex. However, $Mn(CN)_6^{4-}$ is low-spin ($\sim$2.2 B.M. at room temperature). Clearly, caution must be exercised in making conclusions about the magnitude of Δ or B for ligands which bond strongly to metal ions. However, most Mn^{II} complexes do show a magnetic behavior indicative of five unpaired electrons.

NICKEL

An octahedral Ni^{II} complex, $[Ar]3d^8$, should have a $^3A_{2g}$ ground state at all values of Δ/B. Indeed, all known octahedral complexes of Ni^{II} have a ground state which is a spin triplet unless spin–spin coupling occurs between octahedral units in close proximity. In water the labile Ni^{II} undoubtedly has six water molecules about it and exists as the pale green $Ni(H_2O)_6^{2+}$. The electronic spectrum shows spin-allowed transitions at 8500 cm^{-1}, 13,500 cm^{-1}, and 25,300 cm^{-1}, the $^3A_{2g} \rightarrow {}^3T_{2g}$, $^3T_{1g}(F)$, and $^3T_{1g}(P)$ transitions, respectively (see Figure 5-6).

Tetrahedral Ni^{II} species such as $NiCl_4^{2-}$ have been studied in considerable detail. They can be prepared in solvents which are relatively poor bases. Tetrahalonickelate anions are stabilized in the solid state by large cations such as the tetraphenylphosphonium ion. According to the ligand field model, we expect the eight electrons to be distributed among the t_2 and e orbitals as $e^4 t_2^4$. The result is that two "holes" exist in the t_2 level. The Tanabe–Sugano diagram appropriate to this configuration is not cubic d^8 but cubic d^2 [ground state $^3T_1(t_2^2)$]. Spin-allowed transitions are expected to the $^3T_2$, $^3T_1(P)$, and $^3A_2$ levels (Figure 5-3). The value of Δ_t (tetrahedral) is $\sim(4/9)$ Δ_o (octahedral), or as indicated in Table 5-7, $\Delta_t \sim 3800$ cm^{-1} for NiCl$_4^{2-}$ and $\Delta_t/B \sim 4.7$. (From Table 5-7, $\Delta_t = fg = 0.43 \times 8900$ cm$^{-1} \cong 3800$ cm^{-1}. Also, $\beta = B$(complex)$/B$(free ion) $= 1 - hk = 1 - (2.15)(0.12) = 0.742$. Thus B(complex) $\cong 801$ cm^{-1}, since B(free ion) $= 1080$ cm^{-1}. The ratio $\Delta_t/B \cong 4.7$.) Transitions are found near 7800 cm^{-1} ($^3T_1 \rightarrow {}^3A_2$) and 15,000 cm^{-1} ($^3T_1 \rightarrow {}^3T_1(P)$) in the deep blue complex Ni[OAs(C$_6$H$_5$)$_3$]$_2$Cl$_2$ in benzene. A third transition, $^3T_1 \rightarrow {}^3T_2$, is expected near 4000 cm^{-1}. This transition generally is difficult to observe experimentally. Besides being weak in intensity, vibrational transitions also appear in the same spectral region.

The lack of a center of symmetry in tetrahedral complexes causes the absorption intensities of d–d transitions to be considerably larger than for the *centrosymmetric* octahedral species. For example, the visible band for "tetrahedral" Ni[OAs(C$_6$H$_5$)$_3$]$_2$Br$_2$ has a molar extinction, ϵ, of $\sim$170 M^{-1} cm^{-1}, while $\epsilon \sim 3$ for Ni(H$_2$O)$_6^{3+}$ in the same spectral region. The visible spectrometer thus becomes a powerful structural tool.

In the absence of excess coordinating base and when polymerization is prevented by steric features of the ligands (or by sufficient dilution in noncoordinating solvents), some β-ketoenolate complexes such as Ni(DPM)$_2$ (Figure 6-3) are diamagnetic and have a planar NiO$_4$ geometry ($\sim$D$_{2h}$). These red materials readily turn light bluish-green in the presence of adduct-forming water or amines and become paramagnetic. Base adducts of Ni^{II} acetylacetonate are known with *trans* ($\sim$D$_{2h}$), Ni(AcAc)$_2$(C$_5$H$_5$N)$_2$, and *cis* ($\sim$C$_{2v}$), Ni(AcAc)$_2$(C$_5$H$_5$NO)$_2$, geometries. Sulfur ligands tend to produce planar complexes which are often unreactive to excess base.

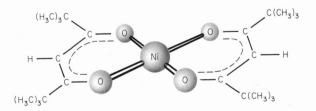

Figure 6-3 Square planar bis(1,1,6,6-tetramethyl-3,5-heptanediono)Ni^{II}, Ni(DPM)$_2$.

COPPER

The relative stability of the Cu^I oxidation state leads to some very important chemistry with copper species. By a one-electron transfer, a Cu^{II} complex may be reduced to Cu^I, leaving a ligand radical. In water, iodide ion reacts with Cu^{II} to produce the precipitate $CuI(s)$ and I_2. However, water soluble Cu^I salts generally disproportionate to the metal and blue $Cu(H_2O)_6^{2+}$. Cu^I compounds are colorless to red-orange, depending on the energy of low-lying electron-transfer (or charge-transfer) transitions. The $[Ar]d^{10}$ configuration rules out the occurrence of d–d electronic transitions. Several Cu^I metal ion clusters have been discovered recently such as the one given in Figure 6-10.

The $[Ar]3d^9$ configuration of Cu^{II} complexes gives the octahedral species a ^2E_g ground state. One visible-region electronic transition is expected, $^2E_g \rightarrow {}^2T_{2g}$, at Δ. However, here again Jahn–Teller effects influence the band shape so that a shoulder is generally observed along with the principal visible absorbance. With Cu^{II} complexes, there appears to be a strong tendency toward tetragonal distortion and the limiting planar configuration. Species such as $CuCl_2$ have a distorted octahedron of ligand atoms about the metal. Generally, considerable elongation along one axis is found, but in the case of K_2CuF_4 two Cu—F distances are substantially shorter than the other four.

JAHN–TELLER BEHAVIOR

A few words about the Jahn–Teller effect seem appropriate, since many transition metal systems are influenced by this phenomenon. We have mentioned that certain complexes, particularly those with E_g ground states, are subject to Jahn–Teller influences. A theorem attributed to the physicists H. A. Jahn and E. Teller states explicitly that *the energy in nonlinear molecules is minimized when distortions occur to remove either spin or orbit degeneracy*, or both. Since the energies associated with molecular distortions due to spin degeneracy appear small compared with distortions caused by orbital effects, we will ignore the influence on spin and illustrate the Jahn–Teller effect for octahedral orbital degeneracy only.

Consider a d^9 ion in an octahedral complex. Using a crystal field model, the half-vacant d orbital may be either $d_{x^2-y^2}$ or d_{z^2} (both e_g) or their linear combination, giving a ^2E_g ground state. If we imagine that the $d_{x^2-y^2}$ orbital contains the "hole" we assume that the in-plane ligand atoms are less effectively shielded from the metal than the two axial ligands and hence bond more strongly to it. An elongated tetragonal complex of D_{4h} symmetry results. The opposite is true when the hole is in d_{z^2}. In reality, however, d_{z^2} and $d_{x^2-y^2}$ are equivalent in O_h and the complex cannot gain any more energy by tetragonal distortion along x than along y or z. However, it will gain energy by a symmetry reduction to D_{4h} with the E_g ground state split into two nondegenerate states (Figure 6-4). The T_{2g} excited state also loses its degeneracy on distortion, but the splitting is much smaller.

Since tetragonal distortion along either x, y, or z is equally feasible, the molecule may undergo a *pseudorotation* or *minima exchange* between the three D_{4h} stereochemistries pictured in Figure 6-5. The average geometry of the complex is octahedral, but its instantaneous

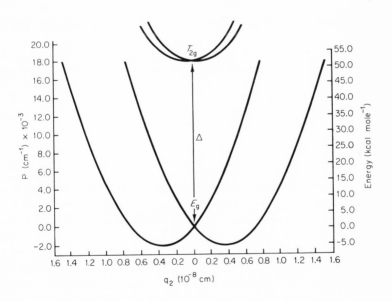

Figure 6-4   *Splitting of E_g and T_{2g} states as q_2, the coordinate describing the Jahn–Teller distortion of an octahedral molecule, is varied. The splitting of T_{2g} is not drawn to scale. [After T. S. Davis, J. P. Fackler, Jr., and M. Weeks, Inorg. Chem. 7, 1994 (1968). By permission of the copyright owner.]*

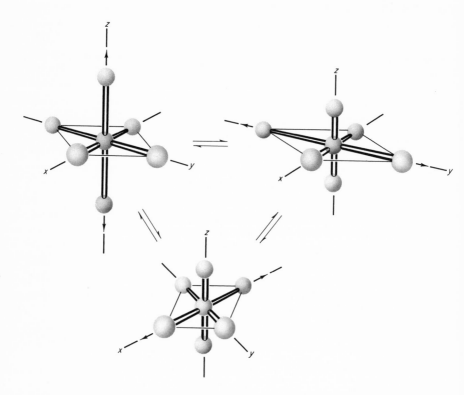

Figure 6-5 *Stereochemical rearrangement or pseudorotation of a tetragonally distorted octahedral complex between the three equally energetic minima.*

geometry is likely to be D_{4h}. Since electronic transitions occur within a time much more rapid than nuclear motions, the electronic spectra may display features consistent with a tetragonal geometry about the metal ion. In solution or in the solid state, molecules cannot undergo nuclear motions without interacting in some way with other molecules. These interactions may produce barriers to the pseudorotation and "freeze" distortions which are then observed by some physical technique such as X-ray crystallography. Thus, the $M(H_2O)_6^{2+}$ ion in $(NH_4)_2M(SO_4)_2 \cdot 6H_2O$ with $M = Cr^{II}$ or Cu^{II} exists as a grossly distorted MO_6 octahedron, while with $M = Ni^{II}$ or Co^{II} (non-E_g ground states) distortions are considerably smaller.

SECOND AND THIRD TRANSITION SERIES

The filling of the 4f shell with electrons to form the lanthanide elements and the resultant increase in nuclear charge causes the 5d orbital probability density in a third transition series element to be concentrated about as close to the nucleus as for a 4d orbital in a second transition series element. The 4d and 5d orbitals thus appear to overlap about equally effectively with ligand orbitals, and ionic sizes of second and third transition series ions in the same triad in a given oxidation state are very similar. In fact, Zr^{IV} and Hf^{IV} are so much alike in their chemistries that it is very difficult to separate compounds of these ions. (The first transition series ion in a triad is much smaller and usually has a much different chemistry.)

The higher oxidation states of the second and third transition series ions tend to be more stable than in the first series. The M^{II} oxidation state is found much less commonly among the heavier elements. In aqueous solution, species such as $M(H_2O)_6^{2+}$ appear non-existent, except perhaps for Cd^{II} and Hg^{II}. Other reasonably stable II ions such as Pd^{II} or Pt^{II} tend to bond strongly with halide or other similar anions which may also be present in solution.

The greater covalency of metal–ligand bonding with compounds of second and third transition series reduces the effectiveness of ligand field theory in describing electronic and magnetic behavior. Spin-orbit coupling (and the breakdown of Russell–Saunders coupling) also strongly influences energy levels, thus making a detailed understanding of the spectra very much less quantitative than is often desirable. Some important attempts have been made to use molecular orbital theory to describe the electronic behavior, particularly with D_{4h} species such as $Pt(CN)_4^{2-}$ or $PdCl_4^{2-}$. While the success achieved is encouraging, it is not quantitative.

NIOBIUM AND TANTALUM

The fluorides and chlorides of pentavalent niobium and tantalum are sufficiently different chemically to be useful in a separation of the metals by crystallization (fluorides) or distillation (chlorides). The elements form complexes having formal oxidation states which vary M^{-I}, as in $M(CO)_6^-$ (O_h symmetry), to M^V, as in MF_7^{2-} ($\sim C_{2v}$ symmetry,

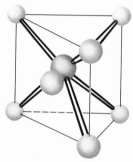

Figure 6-6 *Face-centered trigonal prismatic, seven-coordinate species such as* TaF_7^{2-}
($\sim C_{2v}$).

Figure 6-6). The pentavalent oxides, M_2O_5, are chemically so inert that concentrated acids will not dissolve them unless fluoride is present. With excess fluoride, anions form, such as MF_7^{2-}, TaF_8^{3-}, and TaF_9^{4-}.

The M^{IV} oxidation state is recognized for the unusual metal atom cluster compounds that are formed when the halides are reduced by cadmium or lead and dissolved in aqueous acid. The $Ta_6Cl_{12}^{2+}$ structure is given in Figure 6-7. Among other cluster species are included

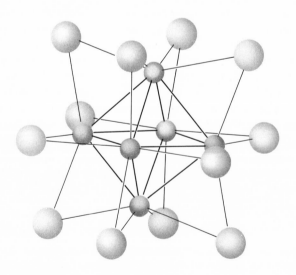

Figure 6-7 *The structure of the complex ion* $[Ta_6Cl_{12}]^{2+}$. *(After L. Pauling as depicted by F. A. Cotton, Accounts Chem. Res.* **2**, *242 (1969).]*

$Nb_6Cl_{14} \cdot 7H_2O$, $Nb_6Br_{14} \cdot 7H_2O$, $Ta_6Cl_{14} \cdot 7H_2O$, and $Ta_6Br_{14} \cdot 7H_2O$. Some reactions of niobium and tantalum and their compounds are

$$NbBr_5(g) + Nb(s) \rightarrow NbBr_4(g)$$

$$2\,TaI_5 + 5\,C_5H_5N \rightarrow 2\,TaI_4(C_5H_5N)_2 + C_5H_5NI_2$$

$$Nb_2O_5 + HF \rightarrow NbOF_5^{2-} \text{ and } NbF_6^-$$

$$MCl_5 + CO \xrightarrow[\text{diglyme}]{\text{Fe(CO)}_5 \text{ catalyst}} M(CO)_6^-$$

$$NaM(CO)_6 + C_5H_5^- + Hg^{2+} \xrightarrow{\text{diglyme}} (\pi\text{-}C_5H_5)M(CO)_4$$

RHENIUM

The reduction of perrhenate, ReO_4^-, in HCl or HBr with hydrogen produces "$ReCl_4^-$," a diamagnetic, dark green species originally thought to be tetrahedral. The electronic configuration for the metal ion in such a complex would be $[Xe]e_g^4$, assuming a large Δ as is commonly found in third transition series complexes. A $^1A_1$ ground state should result for the "spin-paired" tetrahedral species. However, it was found that the diamagnetism of $ReCl_4^-$ was not caused by the T_d symmetry of the anion, but is due to dimerization. The anion $Re_2Cl_8^{2-}$ has the structure presented in Figure 1-2 with $\sim D_{4h}$ symmetry. This is an exceedingly important structure, since it shows how multiple metal–metal bonding can influence the stereochemistry of a complex. F. A. Cotton has shown that the Re—Re bonding in this compound can be described as consisting of a sigma bond, two pi bonds (d_{xz} and d_{yz} orbitals), and one delta bond ($d_{x^2-y^2}$ orbital overlap). This delta bond prevents free rotation about the internuclear Re—Re axis. The very short (2.24 Å) Re—Re distance also is consistent with the high (four) Re—Re bond order.

Other polymeric Re^{III} compounds are known, such as Re_3Cl_9 and Re_3Br_9 (D_{3h} symmetry), $Re_3X_{12}^{3-}$, $X = Cl$ or Br. Oxohalides of Re^{V-VII} are known such as $ReOF_3$, $ReOCl_4$, and $ReOF_5$. They hydrolyze readily in water. Reaction of ReO_4^- with Zn/Hg produces the ReH_9^{3-} ion, the D_{3h} structure of which was presented earlier (Figure 1-12).

TECHNETIUM

This second-row element in the manganese triad is only found as a relatively long-lived radioactive species. The ^{99}Tc isotope emits a β^- (nuclear electron) with a half-life of $\sim 2 \times 10^5$ yr. Since the advent of nuclear reactors to produce the elements, considerable knowledge of technetium chemistry has been gained.

MISCELLANEOUS COORDINATION COMPOUNDS

COMPLEXES CONTAINING MOLECULAR NITROGEN

It is well known that certain plants biochemically synthesize nitrogen-containing compounds from molecular nitrogen in the air. It has been known for several years that the enzymes responsible for the "nitrogen fixation" contain iron and molybdenum, yet until the middle 1960s no nitrogen complexes were known. However, once one was characterized, many examples appeared. The $Ru(NH_3)_5N_2^{2+}$ cation was the first such complex reported. It may be synthesized directly by reacting $Ru(NH_3)_5(H_2O)^{2+}$ with N_2H_4 or by the reaction of $Ru(NH_3)_5N_3^{3+}$ with itself:

$$2\ Ru(NH_3)_5N_3^{2+} \rightarrow 2\ Ru(NH_3)_5N_2^{2+} + 2\ N_2(g)$$

OTHER TRANSITION METAL COMPOUNDS

Structures of coordination compounds containing one transition metal ion (mononuclear) vary from simple gaseous diatomic

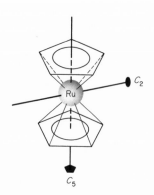

Figure 6-8   Structure of $(\pi\text{-}C_5H_5)_2Ru$, ruthenocene (D_{5h} symmetry).

Table 6-3 *Some mononuclear transition metal complexes*

Species	Symmetry	Species	Symmetry
ScF_6^{3-}	O_h	$ZrCl_4$	T_d (gaseous)
$TiCl_4$	T_d	$NbCl_5$, $TaCl_5$	D_{3h} (gaseous)
$Ti(H_2O)_6^{3+}$	$\sim O_h$	TaF_8^{3-}	$\sim D_{4d}$
VCl_4	$\sim T_d$	$W(CO)_6$	O_h
$VO(H_2O)_5^{2+}$	$\sim C_{4v}$	MoS_2	$\sim D_{3h}$ (6-S atoms)
$Cr(H_2O)_6^{2+}$	$\sim D_{4h}$	ReF_8^-	D_{4d}
$Cr(H_2O)_6^{3+}$	$\sim O_h$	ReO_4^-	T_d
$Cr(CO)_6$	O_h	OsO_4, RuO_4	T_d (gaseous)
MnO_4^-	T_d	$Ru(NH_3)_6^{3+}$	$\sim O_h$
$Fe(CO)_5$	D_{3h}	$(\pi\text{-}C_5H_5)_2Ru$	$\sim D_{5h}$
$FeCl_4^-$	T_d	$Rh(CN)_6^{3-}$, $Rh(H_2O)_6^{3+}$	$\sim O_h$
$Fe(H_2O)_6^{2+}$	$\sim O_h$	P_3RhCl^b	$\sim C_{2v}$ (planar)
$Co(H_2O)_6^{3+}$	$\sim O_h$	$IrCl_6^{3-}$	O_h
$CoCl_4^{2-}$	T_d	$PtCl_4^{2-}$, $Pd(NH_3)_4^{2+}$	$\sim D_{4h}$
$NiCl_4^{2-}$, $Ni(CO)_4$	T_d	PtF_6, PtF_6^-	O_h
$Ni(CN)_5^{3-}$	D_{3h} and C_{4v}	$Ag(CN)_2^-$, $Ag(NH_3)_2^+$	$\sim D_{\infty h}$
$Ni(H_2O)_6^{2+}$	O_h	$Ag(SCN)_4^{3-}$	T_d
NiO_4 (in $Ni(DPM)_2)^a$	$\sim D_{4h}$	$PAuCl_3^b$	$\sim C_{2v}$ (planar)
$CuCl_4^{2-}$	D_{2d} and D_{4h}	$Zn(CH_3)_2$, $Cd(C_6H_5)_2$	$\sim D_{\infty h}$
CuO_4(in $Cu(DPM)_2)^a$	$\sim D_{4h}$	Hg_2Cl_2, Hg_2I_2	$\sim D_{\infty h}$

^{a} DPM = $(CH_3)_3CCOCHCOC(CH_3)_3$.
^{b} P = phosphorus atoms in $(C_6H_5)_3P$.

species ($C_{\infty v}$) to complex structures such as the organometallic species ($\pi\text{-}C_5H_5)_2Ru$ (D_{5h} symmetry, Figure 6-8). In Table 6-3 some important mononuclear complexes of most transition elements are listed along with their molecular symmetries.

CLUSTER COMPOUNDS

Cluster compounds of transition metals are being detected with increasing frequency. Beside the Nb, Ta, and Re species already mentioned, there are numerous others, particularly with ligands containing S or CO. Metal–metal bonding was shown to exist very early in Hg_2Cl_2, but still exists largely in a qualitative sense only for many transition metal compounds. Some metal–metal interaction (as opposed to bond formation) has been postulated for every element in the transition

Table 6-4　*Some polynuclear transition metal complexes*

Compound	Symmetry
$V_{10}O_{28}^{6-}$	(fused VO_6 octahedra)
$Cr_2(O_2CCH_3)_2 \cdot 2H_2O$	$\sim D_{4h}$, acetate bridges
$(OC)_5Mn\text{—}Mn(CO)_5$	D_{4d}
$(OC)_3Fe(CO)_3Fe(CO)_3$	D_{3h}, fused octahedra
$Cl_4NbCl_2NbCl_4$	$\sim D_{2h}$
$Cl_3WCl_3WCl_3^{3-}$	D_{3h}, fused octahedra
$(OC)_2RhCl_2Rh(CO)_2$	$\sim D_{2h}$, planar
$Rh_6(CO)_{16}$	$Rh_6 \sim O_h$, overall $\sim T_d$ with four CO face bridging on the octahedron
$[(CH_3)_3AsCuI]_4$	$\sim T_d$
$(C_3H_5)PdCl_2Pd(C_3H_5)$	$\sim D_{2d}$, pi bonded allyl group

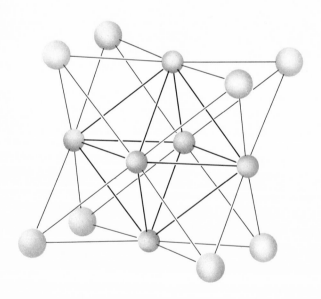

Figure 6-9　*The structure of the* $[Mo_6Cl_8]^{4+}$ *group in* $[Mo_6Cl_8(OH)_4(H_2O)_2 \cdot 12H_2O$. [*After L. Pauling as depicted by F. A. Cotton, Accounts Chem. Res.* **2**, *242 (1969).*]

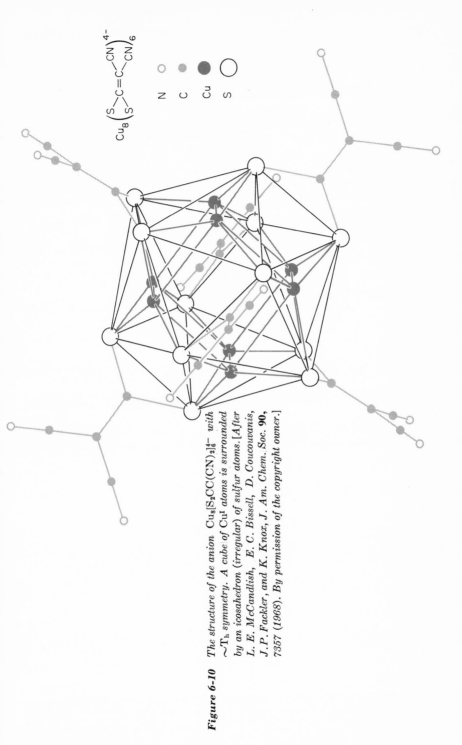

$$Cu_8\left(\begin{matrix}S\\S\end{matrix}C=C\begin{matrix}CN\\CN\end{matrix}\right)_6^{4-}$$

N ○ C ● Cu ● S ○

Figure 6-10 *The structure of the anion* $Cu_8[S_2CC(CN)_2]_6^{4-}$ *with* $\sim T_h$ *symmetry. A cube of* Cu^I *atoms is surrounded by an icosahedron (irregular) of sulfur atoms.* [*After L. E. McCandlish, E. C. Bissell, D. Coucouvanis, J. P. Fackler, and K. Knox, J. Am. Chem. Soc.* **90,** *7357 (1968). By permission of the copyright owner.*]

block excluding only the scandium triad. A very important structure with a metal ion arrangement like that of $Nb_6Cl_{12}^{2+}$ is that of $Mo_6Cl_8^{4+}$ (O_h symmetry, pictured in Figure 6-9). Another very interesting one is the Cu^I anion, $Cu_8[S_2CC(CN)_2]_6^{4-}$ ($\sim T_h$ symmetry) presented in Figure 6-10. The point group symmetries for several additional cluster molecules are presented in Table 6-4.

Some important character tables

APPENDIX

C_1	E
A	1

C_i	E	i		
A_g	1	1	R_x, R_y, R_z	$x^2, y^2, z^2, xy,$ xz, yz
A_u	1	-1	x, y, z	

C_s	E	σ_h		
A'	1	1	x, y, R_z	$x^2, y^2,$ z^2, xy
A''	1	-1	z, R_x, R_y	yz, xz

C_2	E	C_2		
A	1	1	z, R_z	x^2, y^2, z^2, xy
B	1	-1	x, y, R_x, R_y	yz, xz

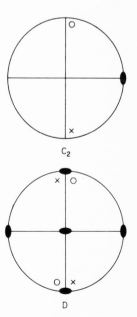

C_2

D

D_2	E	$C_2(z)$	$C_2(y)$	$C_2(x)$		
A	1	1	1	1		x^2, y^2, z^2
B_1	1	1	-1	-1	z, R_z	xy
B_2	1	-1	1	-1	y, R_y	xz
B_3	1	-1	-1	1	x, R_x	yz

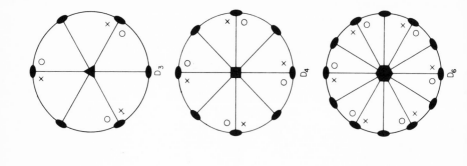

D_3	E	$2C_3$	$3C_2$		
A_1	1	1	1		$x^2 + y^2,\, z^2$
A_2	1	1	-1	$z,\, R_z$	
E	2	-1	0	$(x,\, y)\,(R_x,\, R_y)$	$(x^2 - y^2,\, xy)\,(xz,\, yz)$

D_4	E	$2C_4$	$C_2\,(\equiv C_4^2)$	$2C_2'$	$2C_2''$		
A_1	1	1	1	1	1		$x^2 + y^2,\, z^2$
A_2	1	1	1	-1	-1	$z,\, R_z$	
B_1	1	-1	1	1	-1		$x^2 - y^2$
B_2	1	-1	1	-1	1		xy
E	2	0	-2	0	0	$(x,\, y)\,(R_x,\, R_y)$	$(xz,\, yz)$

D_6	E	$2C_6$	$2C_3$	C_2	$3C_2'$	$3C_2''$		
A_1	1	1	1	1	1	1		$x^2 + y^2,\, z^2$
A_2	1	1	1	1	-1	-1	$z,\, R_z$	
B_1	1	-1	1	-1	1	-1		
B_2	1	-1	1	-1	-1	1		
E_1	2	1	-1	-2	0	0	$(x,\, y)\,(R_x,\, R_y)$	$(xz,\, yz)$
E_2	2	-1	-1	2	0	0		$(x^2 - y^2,\, xy)$

C_{2v}	E	C_2	$\sigma_v(xz)$	$\sigma_v'(yz)$		
A_1	1	1	1	1	z	x^2, y^2, z^2
A_2	1	1	-1	-1	R_z	xy
B_1	1	-1	1	-1	x, R_y	xz
B_2	1	-1	-1	1	y, R_z	yz

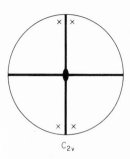

C_{2v}

C_{3v}	E	$2C_3$	$3\sigma_v$		
A_1	1	1	1	z	$x^2 + y^2, z^2$
A_2	1	1	-1	R_z	
E	2	-1	0	$(x, y)(R_x, R_y)$	$(x^2 - y^2, xy)(xz, yz)$

C_{3v}

C_{4v}	E	$2C_4$	C_2	$2\sigma_v$	$2\sigma_d$		
A_1	1	1	1	1	1	z	$x^2 + y^2, z^2$
A_2	1	1	1	-1	-1	R_z	
B_1	1	-1	1	1	-1		$x^2 - y^2$
B_2	1	-1	1	-1	1		xy
E	2	0	-2	0	0	$(x, y)(R_x, R_y)$	(xz, yz)

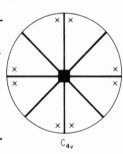

C_{4v}

C_{2h}	E	C_2	i	σ_h		
A_g	1	1	1	1	R_z	x^2, y^2, z^2, xy
B_g	1	-1	1	-1	R_x, R_y	xz, yz
A_u	1	1	-1	-1	z	
B_u	1	-1	-1	1	x, y	

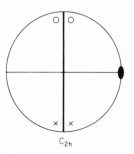

C_{2h}

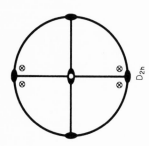

D_{2h}

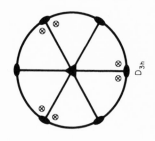

D_{3h}

D_{2h}	E	$C_2(z)$	$C_2(y)$	$C_2(x)$	i	$\sigma(xy)$	$\sigma(xz)$	$\sigma(yz)$		
A_g	1	1	1	1	1	1	1	1		x^2, y^2, z^2
B_{1g}	1	1	-1	-1	1	1	-1	-1	R_z	xy
B_{2g}	1	-1	1	-1	1	-1	1	-1	R_y	xz
B_{3g}	1	-1	-1	1	1	-1	-1	1	R_x	yz
A_u	1	1	1	1	-1	-1	-1	-1		
B_{1u}	1	1	-1	-1	-1	-1	1	1	z	
B_{2u}	1	-1	1	-1	-1	1	-1	1	y	
B_{3u}	1	-1	-1	1	-1	1	1	-1	x	

D_{3h}	E	$2C_3$	$3C_2$	σ_h	$2S_3$	$3\sigma_v$		
A_1'	1	1	1	1	1	1		$x^2 + y^2, z^2$
A_2'	1	1	-1	1	1	-1	R_z	
E'	2	-1	0	2	-1	0	(x, y)	$(x^2 - y^2, xy)$
A_1''	1	1	1	-1	-1	-1		
A_2''	1	1	-1	-1	-1	1	z	
E''	2	-1	0	-2	1	0	(R_x, R_y)	(xz, yz)

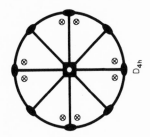

D_{4h}

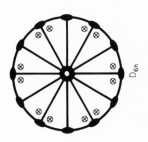

D_{6h}

D_{4h}	E	$2C_4$	C_2	$2C_2'$	$2C_2''$	i	$2S_4$	σ_h	$2\sigma_v$	$2\sigma_d$		
A_{1g}	1	1	1	1	1	1	1	1	1	1		$x^2+y^2,\ z^2$
A_{2g}	1	1	1	-1	-1	1	1	1	-1	-1	R_z	
B_{1g}	1	-1	1	1	-1	1	-1	1	1	-1		x^2-y^2
B_{2g}	1	-1	1	-1	1	1	-1	1	-1	1		xy
E_g	2	0	-2	0	0	2	0	-2	0	0	(R_x, R_y)	(xz, yz)
A_{1u}	1	1	1	1	1	-1	-1	-1	-1	-1		
A_{2u}	1	1	1	-1	-1	-1	-1	-1	1	1	z	
B_{1u}	1	-1	1	1	-1	-1	1	-1	-1	1		
B_{2u}	1	-1	1	-1	1	-1	1	-1	1	-1		
E_u	2	0	-2	0	0	-2	0	2	0	0	(x, y)	

D_{6h}	E	$2C_6$	$2C_3$	C_2	$3C_2'$	$3C_2''$	i	$2S_3$	$2S_6$	σ_h	$3\sigma_d$	$3\sigma_v$		
A_{1g}	1	1	1	1	1	1	1	1	1	1	1	1		$x^2+y^2,\ z^2$
A_{2g}	1	1	1	1	-1	-1	1	1	1	1	-1	-1	R_z	
B_{1g}	1	-1	1	-1	1	-1	1	-1	1	-1	1	-1		
B_{2g}	1	-1	1	-1	-1	1	1	-1	1	-1	-1	1		
E_{1g}	2	1	-1	-2	0	0	2	1	-1	-2	0	0	(R_x, R_y)	(xz, yz)
E_{2g}	2	-1	-1	2	0	0	2	-1	-1	2	0	0		$(x^2-y^2,\ xy)$
A_{1u}	1	1	1	1	1	1	-1	-1	-1	-1	-1	-1		
A_{2u}	1	1	1	1	-1	-1	-1	-1	-1	-1	1	1	z	
B_{1u}	1	-1	1	-1	1	-1	-1	1	-1	1	-1	1		
B_{2u}	1	-1	1	-1	-1	1	-1	1	-1	1	1	-1		
E_{1u}	2	1	-1	-2	0	0	-2	-1	1	2	0	0	(x, y)	
E_{2u}	2	-1	-1	2	0	0	-2	1	1	-2	0	0		

D$_{2d}$

D$_{3d}$

D$_{2d}$	E	$2S_4$	C_2	$2C_2'$	$2\sigma_{\rm d}$		
A_1	1	1	1	1	1		$x^2 + y^2,\ z^2$
A_2	1	1	1	-1	-1	R_z	
B_1	1	-1	1	1	-1		$x^2 - y^2$
B_2	1	-1	1	-1	1	z	xy
E	2	0	-2	0	0	(x, y); (R_x, R_y)	(xz, yz)

D$_{3d}$	E	$2C_3$	$3C_2$	i	$2S_6$	$3\sigma_{\rm d}$		
A_{1g}	1	1	1	1	1	1		$x^2 + y^2,\ z^2$
A_{2g}	1	1	-1	1	1	-1	R_z	
E_g	2	-1	0	2	-1	0	(R_x, R_y)	$(x^2 - y^2,\ xy)$, (xz, yz)
A_{1u}	1	1	1	-1	-1	-1		
A_{2u}	1	1	-1	-1	-1	1	z	
E_u	2	-1	0	-2	1	0	(x, y)	

T$_{\rm d}$	E	$8C_3$	$3C_2$	$6S_4$	$6\sigma_{\rm d}$		
A_1	1	1	1	1	1		$x^2 + y^2 + z^2$
A_2	1	1	1	-1	-1		
E	2	-1	2	0	0		$(2z^2 - x^2 - y^2,$ $x^2 - y^2)$
T_1	3	0	-1	1	-1	(R_x, R_y, R_z)	
T_2	3	0	-1	-1	1	(x, y, z)	(xy, xz, yz)

O_h	E	$8C_3$	$6C_2$	$6C_4$	$3C_2(\equiv C_4^2)$	i	$8S_6$	$6\sigma_d$	$6S_4$	$3\sigma_h$		
A_{1g}	1	1	1	1	1	1	1	1	1	1		$x^2+y^2+z^2$
A_{2g}	1	1	-1	-1	1	1	1	-1	-1	1		
E_g	2	-1	0	0	2	2	-1	0	0	2		$(2z^2-x^2-y^2,\ x^2-y^2)$
T_{1g}	3	0	-1	1	-1	3	0	-1	1	-1	(R_x, R_y, R_z)	
T_{2g}	3	0	1	-1	-1	3	0	1	-1	-1		(xz, yz, xy)
A_{1u}	1	1	1	1	1	-1	-1	-1	-1	-1		
A_{2u}	1	1	-1	-1	1	-1	-1	1	1	-1		
E_u	2	-1	0	0	2	-2	1	0	0	-2		
T_{1u}	3	0	-1	1	-1	-3	0	1	-1	1	(x, y, z)	
T_{2u}	3	0	1	-1	-1	-3	0	-1	1	1		

I_h	E	$12C_5$	$12C_5^2$	$20C_3$	$15C_2$	i	$12S_{10}$	$12S_{10}^3$	$20S_6$	15σ		
A_g	1	1	1	1	1	1	1	1	1	1		$x^2+y^2+z^2$
T_{1g}	3	$\frac{1}{2}(1+\sqrt{5})$	$\frac{1}{2}(1-\sqrt{5})$	0	-1	3	$\frac{1}{2}(1-\sqrt{5})$	$\frac{1}{2}(1+\sqrt{5})$	0	-1	(R_x, R_y, R_z)	
T_{2g}	3	$\frac{1}{2}(1-\sqrt{5})$	$\frac{1}{2}(1+\sqrt{5})$	0	-1	3	$\frac{1}{2}(1+\sqrt{5})$	$\frac{1}{2}(1-\sqrt{5})$	0	-1		
G_g	4	-1	-1	1	0	4	-1	-1	1	0		
H_g	5	0	0	-1	1	5	0	0	-1	1		$(2z^2-x^2-y^2,$ $x^2-y^2,$ $xy,\ yz,\ zx)$
A_u	1	1	1	1	1	-1	-1	-1	-1	-1		
T_{1u}	3	$\frac{1}{2}(1+\sqrt{5})$	$\frac{1}{2}(1-\sqrt{5})$	0	-1	-3	$-\frac{1}{2}(1-\sqrt{5})$	$-\frac{1}{2}(1+\sqrt{5})$	0	1	(x, y, z)	
T_{2u}	3	$\frac{1}{2}(1-\sqrt{5})$	$\frac{1}{2}(1+\sqrt{5})$	0	-1	-3	$-\frac{1}{2}(1+\sqrt{5})$	$-\frac{1}{2}(1-\sqrt{5})$	0	1		
G_u	4	-1	-1	1	0	-4	1	1	-1	0		
H_u	5	0	0	-1	1	-5	0	0	1	-1		

Subject index